# Microsoft Conversational AI Platform for Developers

## End-to-End Chatbot Development from Planning to Deployment

Stephan Bisser

Apress®

*Microsoft Conversational AI Platform for Developers: End-to-End Chatbot Development from Planning to Deployment*

Stephan Bisser
Gratwein, Austria

ISBN-13 (pbk): 978-1-4842-6836-0
https://doi.org/10.1007/978-1-4842-6837-7

ISBN-13 (electronic): 978-1-4842-6837-7

Managing Director, Apress Media LLC: Welmoed Spahr
Acquisitions Editor: Joan Murray
Development Editor: Laura Berendson
Coordinating Editor: Jill Balzano

Cover image designed by Freepik (www.freepik.com)

Distributed to the book trade worldwide by Springer Science+Business Media LLC, 1 New York Plaza, Suite 4600, New York, NY 10004. Phone 1-800-SPRINGER, fax (201) 348-4505, e-mail orders-ny@springer-sbm. com, or visit www.springeronline.com. Apress Media, LLC is a California LLC and the sole member (owner) is Springer Science + Business Media Finance Inc (SSBM Finance Inc). SSBM Finance Inc is a **Delaware** corporation.

For information on translations, please e-mail booktranslations@springernature.com; for reprint, paperback, or audio rights, please e-mail bookpermissions@springernature.com.

Apress titles may be purchased in bulk for academic, corporate, or promotional use. eBook versions and licenses are also available for most titles. For more information, reference our Print and eBook Bulk Sales web page at http://www.apress.com/bulk-sales.

Any source code or other supplementary material referenced by the author in this book is available to readers on GitHub via the book's product page, located at www.apress.com/9781484268360. For more detailed information, please visit http://www.apress.com/source-code.

Printed on acid-free paper

*I dedicate this book to my family, especially to my wife Nicole,
my two beautiful kids Nele and Diego,
to my father whom I miss very much, and to my mother.
Ich liebe euch!*

# Table of Contents

# About the Author

**Stephan Bisser** is a technical lead at Solvion and a Microsoft MVP for artificial intelligence based in Austria. In his current role, he focuses on conversational AI, Microsoft 365, and Azure. He is passionate about the conversational AI platform and the entire Microsoft Bot Framework and Azure Cognitive Services ecosystem. Stephan and several other MVPs founded the Bot Builder Community, which is a community initiative helping Bot Framework developers with code samples and extensions. Together with Thomy Gölles, Rick Van Rousselt, and Albert-Jan Schot, Stephan is hosting SelectedTech, where they publish webinars and videos on social media around SharePoint, Office 365, and the Microsoft AI ecosystem. In addition, he blogs regularly and is a contributing author to *Microsoft AI MVP Book*.

He can be reached at his blog (`https://bisser.io`), at Twitter (@stephanbisser), on LinkedIn (`www.linkedin.com/in/stephan-bisser/`), and on GitHub (stephanbisser).

# About the Technical Reviewer

**Fabio Claudio Ferracchiati** is a senior consultant and a senior analyst/developer using Microsoft technologies. He is a Microsoft Certified Solution Developer for .NET, a Microsoft Certified Application Developer for .NET, a Microsoft Certified Professional, and a prolific author and technical reviewer. Over the past ten years, he's written articles for Italian and international magazines and coauthored more than ten books on a variety of computer topics.

# Acknowledgments

I want to thank my employer, **Solvion**, for being the best company making a lot of things possible which seemed to be impossible a while ago. Of course, I want to thank every colleague as our work-friendship is special. But I need to thank you, **Thomy**, in particular, as it is an honor to work together with you, not only on company topics but also on the community bits and pieces. With that I also may thank the whole MVP and non-MVP community, especially my friends **Rick** and **Appie** who keep me busy with new ideas and projects!

Thank you, **Herbert** and **Brigitte**, for being the best in-laws I could have wished for!

I want to give thanks to my dad for teaching me everything he knew, which is the reason why I am who I am today! *Ich vermisse dich*, **Papa**.

Thank you, Mama, for being here for me in my life and supporting me wherever you can! *Danke*, **Mama**!

I want to thank my two beautiful kids, **Diego** and **Nele**, for keeping me fit and bringing incredibly much joy into my life. Without you I would miss something. *Ich liebe euch beide!*

And lastly, I want to say thank you to my wife, **Nicole**, for supporting me no matter what. I know that it is sometimes hard to be with me, but I could not be happier to have you by my side. You are the best! *Ich liebe dich!*

# Introduction

This book covers all steps which you will encounter when building a chatbot using the Microsoft conversational AI platform. You will learn the most important facts and concepts about the Microsoft Bot Framework and Azure Cognitive Services, which are needed to develop and maintain a chatbot. This book is mainly targeting developers and people who have a development background, as you will learn the concepts of modern end-to-end chatbot development. But as it covers the basic concepts as well, nearly every IT-savvy person who is interested in chatbot development can benefit from reading this book.

Starting with the theoretical concepts, the first three chapters will give detailed insights into the Microsoft conversational AI platform, the Microsoft Bot Framework, and Azure Cognitive Services, which are the services and platforms used throughout this book when it is about developing chatbots. Chapters 4–8 then cover the various stages of every chatbot development project, starting with the design phase where design principles are discussed, followed by the build stage where a real-world example chatbot will be developed using Microsoft Bot Framework Composer, which is a visual bot editing tool. After that, the test phase will be addressed to see how to properly test a chatbot before the last two chapters that will be dealing with the publish and the connect phase in order to offer the chatbot to end users.

**CHAPTER 1**

# Introduction to the Microsoft Conversational AI Platform

Conversational AI (short for artificial intelligence) is a term which can be described as a subfield or discipline of artificial intelligence in general. This area deals with the question on how to expose software or services through a conversational interface, enhanced by AI algorithms. Those conversational interfaces usually range from simple text-based chat interfaces to rather complex voice and speech interfaces. The primary goal is to let people interact with software services through those conversational interfaces to make the interaction between humans and machines more natural. An example of that would be the following:Imagine you sit in your car and want to regulate the heating. In the present you would likely need to push a button or turn a switch within your car to turn the heating either up or down. In a more modern scenario, it would be far more natural to just say something like "Car, please turn the heating to 22 degrees Celsius" to achieve the same. In that second case, you would not need to push a button which distracts you from driving; you would only need to speak and tell the device what it should do. Therefore, you would not only have the possibility to stay focused on the main thing you do which is driving, but this could also be beneficial as it may be faster to say something rather than to control a system via buttons or switches.

This first chapter of the book will outline the key concepts and principles of conversational AI focusing on the Microsoft ecosystem, along with usage scenarios for conversational AI services.

© Stephan Bisser 2021
S. Bisser, *Microsoft Conversational AI Platform for Developers*, https://doi.org/10.1007/978-1-4842-6837-7_1

# Key Concepts of Conversational AI

When talking about "Conversational AI," you always need to think about an orchestration of various components. Those components need to be aligned to deliver the best possible conversational interface for the people using it. In many cases, a comprehensive conversational AI interface may consist of the following:

- Natural language processing (NLP)

- Text analytics

- Text-to-speech

- Speech-to-text

- Text/speech translation

But why is that so important nowadays? The answer is rather simple if you look at what people were using in the past. Figure 1-1 outlines that there is a bit of a paradigm shift happening currently, as it's more important than ever to have a conversation with a service rather than "just" clicking and browsing through.

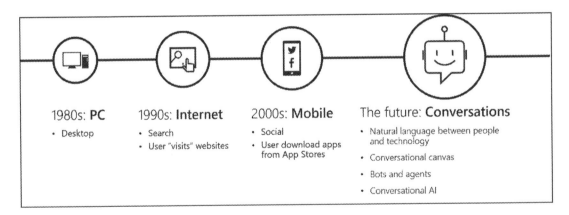

*Figure 1-1.* *Evolution of IT*

Many decades ago, when PCs became popular, the main use case for personal computers has been to work on different tasks on your own, without having the ability to connect with other people via your computer. With the introduction of the Internet, this evolved a bit, as you then could connect your computer with others. This led to the possibility for every user to interact with others via the Internet. In the 2000s, the mobile era began, where people started using smartphones and social media apps to connect

2

via their mobile devices with others. From that point on, people did not need to use a personal computer or laptop anymore to interact with other people. You could just use your smartphone and start a chat over various social media channels with others and have a conversation with them. But until then, it has always been about the interaction between one person and another using different communication channels and services.

This is now slightly changing as there is now a new era starting, where the main focus is on the ability of humans having conversations not only with other people but also with the services they are using in a human way. Imagine a rather simple but quite unusual experience if you would not interact with your coffee machine as you are used to right now, clicking buttons to make yourself a coffee. With the new approach, it could be possible to talk to your coffee machine and say, "I want to have a cappuccino. Could you please make one?", instead. And this approach could also be used with software services as you use them right now, but instead of clicking buttons here and there to tell the service what it should do, you just have a natural conversation with that said service, as you would with another human being. To some extent this is a new sort of user interface. This new UI needs to be intuitive of course and therefore relies on capabilities to understand the user.

# Natural Language Processing

With this new approach of consuming and interacting with services in a more natural or human way, the ability of processing natural language is crucial. And this is one of the many fields in conversational AI which should involve not only developers and engineers but also domain matter experts. It doesn't make sense that you as a developer building a conversational AI app, like a bot, construct the language model for your app. The reason for that is that in most cases, you are not the domain matter expert your app should be developed for. Therefore, it makes sense to collaborate on those fields, like language models, with domain matter experts and the people who know the target audience of your bot. Those domain matter experts have in many cases a better understanding of the way users may interact with your bot and can therefore increase the quality of your cognitive skills within your bot. The key factors of such a natural language understanding model, which is a key component in every conversational AI app, are intents and utterances.

Intents are different areas within your language understanding model. Those intents can be used to indicate what the user's intention is when telling the bot something. They can be used as indicators or triggers within your business logic of an app, to define how the bot should respond to a given input. For example, if a user says, "What's the weather today?" you certainly want the bot respond with something like "It will be mostly sunny with 28 degrees Celsius." To define such an intent, like in the example "GetWeather," you need to provide the natural language model with example phrases, which are called utterances. These utterances for the given intent could be

- What is the weather like?

- How is the weather?

- What's the weather?

- Could you tell me the weather forecast?

- I would like to get some details on the weather please.

As you can see, these five example phrases differ in phrasing and syntax. And this is crucial for the quality level of your language model, as you need to empathize on the end users and how they will ask your bot about the weather. And the more utterances you will provide your model with, the more possibilities your model will cover.

But there is even more to that, as this basic example shows. Now with this scenario, user asks X and bot responds with Y. But you also need to somehow take the context of that conversation into consideration. Because if a user would ask a bot only the question about the weather, the bot does not know the location for which the user wants to know the weather. So, what you would need to add to your language model are so-called entities to indicate certain properties you need to know before you can process the request. In this scenario the location would be such an entity you would need to know before you can respond with details on the weather. To have this ability, you would need to enhance your language understanding model by adding the following utterances and tagging the entities, for example:

- What is the weather like in **{location=Seattle}**?

- How is the weather in **{location=New York}**?

- What's the weather in **{location=Berlin}**?

- Could you tell me the weather forecast for **{location=Redmond}**?

- I would like to get some details on the weather in **{location=Amsterdam}** please.

The language understanding model is now able to detect location entities within given phrases and tag those locations for you. Within the business logic of your app, you can now use those entities, like variables, and process the request accordingly. Now it would be possible that you respond with the correct weather details if a user asks, "What's the weather today in Seattle?" And what you would need to handle within your business logic is the process if a user does not provide a location for querying the weather. In most cases it makes sense to respond with a question stating, "In order to give you the weather forecast, please tell me city you would like to know the weather for." Now this makes the interaction between the user and the bot more natural as there is some sort of context awareness.

But natural language processing is only one component within a conversational AI application. Imagine that you want to build a chatbot which extends your customer service scenario to provide users an additional contact which can be reached 24/7. Now in many cases, a chatbot could be a good way of handling questions or inquiries in a first step to relieve customer service employees. A chatbot can easily handle the basic questions, but what if a customer or end user has an urgent inquiry which cannot be handled by the bot? Analyzing the customer's messages to derive meaning from that can help you routing the messages to the right contact person in the back end. Therefore, it makes sense to use some sort of text analytics engine, which can detect the sentiment from messages. This way, you can decide within your chatbot's business logic if it is necessary to hand off to a human instead of continuing the conversation with the bot.

# Language Translation

Let's extend the preceding use case with handling flight bookings, which can be handled by the bot. Now in many cases, the bot can help users with their bookings. But you would also need to take complaints into consideration. Even that can be managed by a bot; however, many people would like to get a clear response immediately if they want to issue a complaint. Using text analytics of some sort, you could detect the user's sentiment and decide based on the sentiment score, if it is necessary to immediately escalate the conversation to a customer service agent as the customer is angry or upset. This could prevent the case of unsatisfied customers which would lead to customer loss, as they would still be served accordingly based on their situation. Additionally, customer service agents would have more time to work on the "hard" cases and can serve customers more efficiently, as basic questions will be handled by the bot completely.

Furthermore, if you are building a bot which should deal with customer cases, depending on your offerings, it could be also beneficial to implement a multilingual bot. Now there are two possible implementation approaches:

- Build separate bots for each language.

- Build one bot and use translation to enable multilingualism.

Of course, it depends on the scenario and other circumstances like budget and timelines, but in many cases the second approach could be a good alternative to implement a multilingual bot. Using translation could enable your employees responsible for the bot's content and knowledge to only manage the content in one language instead of building knowledge bases or data silos for each language the bot should be using. The high-level message flow for such a multilingual bot could be as illustrated in Figure 1-2.

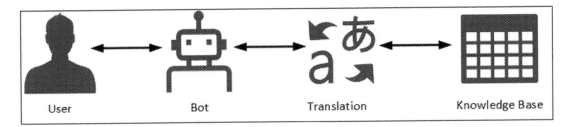

***Figure 1-2.***  *Bot translation flow – high level*

## Speech

Another angle which should not be neglected as well is the question if the bot should be a text-only bot or if it should also be voice-enabled. Voice is certainly a more sophisticated discipline when it comes to bot design. The reason for that is many people are used to writing with other people or bots without using accent phrases. But when people talk, they are used to talk with an accent instead of the grammatically correct way as if they would write something. Therefore, when you are building a bot which should be capable of having a conversation via speech, you would need to think of all the scenarios and accents people might use within their conversations. In addition, what you would need to figure it is the type of architecture you want to implement. For bots, especially in the Microsoft conversational AI platform, a solid scenario is to use

Azure Speech Services, which is a service targeting voice scenarios for implementing speech-to-text and text-to-speech. In such a case, you could use speech-to-text to gather all incoming voice messages and transform them into text, before redirecting those text messages to your bot's logic for executing the business logic. When the bot sends a response back to the user, you could use text-to-speech to translate that text-based message again into a voice-based message before sending it to the user's channel. This kind of architecture allows you to develop a bot which is capable of answering messages in a text-based scenario like chat as well as in a voice-based scenario like a phone call with the same business logic in the back end. The only difference from a development aspect would be to treat the user's channels differently to make sure that each call will use speech-to-text and text-to-speech before doing the natural language processing routines. With that in mind, a major advantage is that you can develop your bot once and then roll it out to text-based and voice-based channels at the same time without changing the code or needing to develop two different bots.

# Usage Scenarios of Conversational AI

There are many different scenarios in which conversational AI can be useful. In fact, one could solve almost every business problem using AI of some sort. The major risk nowadays is to forget about the process of requirements engineering when starting to build conversational AI applications. Conversational AI is a subset of artificial intelligence, focusing on building seamless conversations between computers and humans. But as it is a big buzzword nowadays, the right use cases are often not evaluated thoroughly. Many companies and vendors try to implement AI services because it is a trend these days. But this often results in a situation where users do not benefit from the new solutions powered by AI. The key point which makes an AI-powered use case, especially one built using conversational AI, successful is to make the user experience significantly better or faster and not worse.

One simple example could be an intranet scenario in a large enterprise. Typically, those intranet solutions are designed to store a lot of information like documents in it. The problem with that is that in many cases, those solutions grow over many years and users are having trouble to find relevant information in that "information jungle," which is especially true for new employees using the intranet for the first time. If you imagine now which steps a user typically would need to do to find the right information, this process could look like the following:

1. Log in to the intranet solution.

2. Navigate to the correct area within the intranet where the information might be stored.

3. Navigate to the document store within that area.

4. Look through the documents to find the correct document.

5. Open the document.

Now imagine the same use case but served through a conversational AI app in the form of a chatbot. The users would typically need to do the following to achieve the same result:

1. Log in to the channel where the bot can be accessed.

2. Send a message like: "Could you please send me the document xyz" or "Could you please send me all documents with the metadata xyz."

3. Open the documents sent by the chatbot.

Now if you compare those two approaches, the user would be much faster to just tell the chatbot which information he wants to get, without the necessity of searching through the files himself. This in turn has the potential to save the users a lot of time, which they can invest in innovative tasks or tasks which they like to do as they are more enjoyable other than searching through documents. Of course, this depends on an architecture, in which the bot is integrated in the intranet solution in that case, being able to index all the documents, to send them out to users in a matter of seconds. But the key point here is that the bot or, in general, the applied AI solutions generate value for the end users instead of making the process worse.

Another example would be in the field of business reporting. There are a lot of solutions available in the market which offer services for different reporting purposes. The big problem with many of those is the complexity for the end users in most cases. As a reporting tool should be able to deal with complex data structures to build sophisticated reports, the user interface to build those reports is also quite complex. It would be far easier if you could tell the service which KPIs you want to see using natural language. Therefore, many vendors, including Microsoft, are building services into their reporting solutions, which offer the possibility to query the data using language instead of the complex reporting interface, which they are used to. This way, the end users potentially save a lot of time, as they only need to tell the system what they want to see which usually is a matter of seconds, instead of spending a lot of time in building dashboards upfront. Therefore, conversational AI can be seen as some sort of new user interface, designed to expose software services through a conversational interface, which should be relatively easy to handle as humans are used to speak and interact using natural language.

Another key point of conversational AI is that, by introducing some sort of conversational AI in organizations, those services potentially could take away people's fear of artificial intelligence. In many parts of the world, especially in manufacturing companies, people have some sort of fear that AI services, or generally spoken machines, will take over their job sooner or later. Therefore, they refuse to use any of those "modern" solutions as they think that the machine will take their job and they would face serious problems. By introducing "lightweight" AI services, in the form of chatbots and other conversational AI solutions, companies have the chance to tell people that these will be acting as virtual assistants for their employees, designed to help and assist people. Those virtual assistants should be designed to handle repetitive tasks of users, so that they have the chance to spend more time on innovative or harder tasks, which virtual assistants could not accomplish. By following such an approach, adoption rates of conversational AI services in organizations are typically much higher, if people see those services as assistants designed to take over work, which they do not like to do.

One good example for such a scenario would be in IT support. Every company has some sort of helpdesk team, which basically deals with a variety of different questions and problems. One of the most common problems is the question on the password reset. Imagine that you would work in an IT helpdesk and get the question how to reset a password several times a day. At some point, this could be an annoying question, but still you would need to help users struggling with an expired or forgotten password. But if you would introduce an IT helpdesk chatbot, which would be able to deal with those basic questions like how to reset a user's password, you would have more time to work on the more sophisticated problems, or even work on the process optimization within the support unit. And as the chatbot will hardly ever tell you that he is annoyed of answering the same questions over and over again, the users will be served properly, if not more efficient, as the chatbot can immediately deliver the answer to the user having the problem, as he is always available, even outside of business hours.

# Service Offerings Around Conversational AI in Microsoft Azure

As Microsoft's public cloud platform, Azure offers many different services and solutions for the different areas like Infrastructure as a Service (IaaS), Platform as a Service (PaaS), and Software as a Service (SaaS); it also offers many services in the field of AI as a Service (AIaaS). The service range is quite broad, covering many different service types used in a conversational AI application, which can be seen in Figure 1-3.

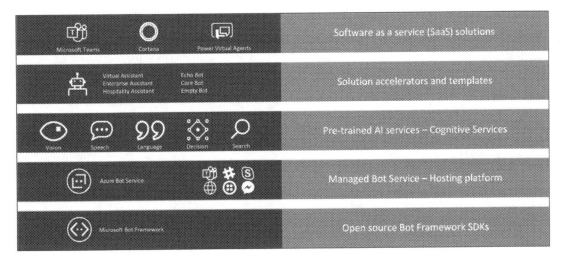

**Figure 1-3.** *Microsoft conversational AI platform*

# Bot Framework and Azure Bot Service

The foundation of every conversational AI application or chatbot is the Microsoft Bot Framework. This is an open-source SDK, which is by now available in C#, JavaScript, and Python (all generally available) as well as Java (Preview). This SDK offers an open, modular, and extensible architecture for building bots and conversational AI apps. It provides a built-in dialog system which is agile and can be customized and adapted. One of the main advantages of this SDK is the fact that it shares the same implementation across all the development languages. This means that developers can build bots using .NET and JavaScript and do not need to learn two different implementation approaches or concepts as they are unified across all programming languages. Moreover, the Microsoft Bot Framework SDK integrates seamlessly into the Microsoft Azure platform as well as all the Microsoft Cognitive Services.

When deploying a bot built using the Microsoft Bot Framework, the Azure Bot Service acts as the hosting platform in Azure. This service basically accelerates the deployment and management cycles of a Bot Framework bot. The Azure Bot Service (ABS) allows us to connect a bot with multiple channels in a matter of minutes, as Microsoft is managing the integration between a bot and the connected channels for us. So you do not need to spend days setting up a new connection between your Azure environment and Facebook, for example, just to activate your bot in the Facebook Messenger – this is done by Microsoft. All you need to care about is that your bot's logic and code are supported by the channel's platform. And a big upside of this is the fact that Microsoft is constantly updating the list of supported channels. By the time of writing, the following channels are supported (with subject to change):

- Cortana
- Office 365 email
- Microsoft Teams
- Skype
- Slack
- Twilio (SMS)
- Facebook Messenger
- Kik Messenger
- GroupMe
- Facebook for Workplace
- LINE
- Telegram
- Web Chat
- Direct Line
- Direct Line Speech

As the preceding list shows, Microsoft is also integrating third-party platforms like Facebook Messenger or Slack. This extends the possible chatbot use cases as the bot can be used in Slack and Microsoft Teams simultaneously sharing the same code base.

Of course, you would need to make sure that the bot's code is supported in both platforms, especially when dealing with front-end matters, because, for example, Microsoft Teams is supporting a concept called Adaptive Cards, which is explained in a later chapter. This solution offers the possibility to render rich attachments like images, audio files, buttons, and texts in one single message. The upside of this is that the definition of such an Adaptive Card is done using JSON and can therefore be replaced in runtime. The downside is that many channels do not yet support this new concept, like Slack or Facebook Messenger. So, if you build a bot for Teams and Slack, you will need to treat both channels differently in different areas within your conversation. But the overall goal is to write a bot once and deploy it in multiple channels. This way you can minimize your development and deployment cycles and maximize the number of possible users and the reach. The details of the Bot Framework and all its concepts and patterns are explained in a later chapter.

# Cognitive Services

Microsoft offers a set of prebuilt machine learning services called "Cognitive Services" which should help developers, without data scientist skills, to build intelligent applications. This set of comprehensive APIs is designed to deliver the ability to develop an application that can decide, understand, speak, hear, and search without a huge amount of effort. The following list summarizes all Cognitive Services which are currently available:

- **Decision**
  - **Anomaly Detector (preview)**
    - A service helping users to foresee any kind of problem before it occurs.
  - **Content Moderator**
    - An API which can be used to moderate content using machine assistance as well as a human review tool for images, text, and videos.
  - **Personalizer**
    - A service based on reinforcement learning to provide personalized and tailored content for end users.

- **Language**

  - **Immersive Reader (preview)**

    - A service which is currently in preview providing the ability to embed text reading and comprehension functionality into applications.

  - **Language Understanding**

    - Integrate natural language understanding functionality based on machine learning models into applications as well as bots to understand end users and derive actions based on their needs.

  - **QnA Maker**

    - A cloud-based service giving developers and business users to build knowledge bases based on frequently asked questions to be used in a conversational manner.

  - **Text Analytics**

    - Text Analytics empowers users to integrate sentiment and language detection, entity, and key phrase extraction from an input text into apps and bots.

  - **Translator**

    - A neural machine translation API for developers offering easy-to-use interfaces to conduct real-time text translation supporting more than 60 languages.

- **Speech**

  - **Speech to Text (part of the speech service)**

    - A real-time speech-to-text API giving users the ability to convert spoken audio into text including the ability to use custom vocabularies to overcome speech recognition barriers.

- **Text to Speech (part of the speech service)**

  - Using neural text-to-speech capabilities, this API offers the ability to convert text-to-speech for natural conversational interfaces supporting a broad range of languages, voices, and accents.

- **Speech Translation (part of the speech service)**

  - A cloud-based automatic translation API offering speech translations in real time.

- **Speaker Recognition (preview)**

  - Speaker Recognition is a service providing the functionality to recognize and identify individual speakers to either determine the identity of an unknown speaker or use speech as a verification method.

- **Vision**

  - **Computer Vision**

    - An API which can analyze the content in images as well as extract text from images and recognize familiar objects in images like brands or landmarks.

  - **Custom Vision**

    - Based on the Computer Vision service, it enables Custom Vision users to create their own computer vision models based on their requirements.

  - **Face**

    - The Face API is tailored to analyze faces in images to be used as facial recognition service which can be integrated in any kind of application.

  - **Form Recognizer (preview)**

    - This API is an AI-powered document extraction solution used to automate the extraction of information like text, key-value pairs, or tables from documents.

- **Ink Recognizer (preview)**

  - Ink Recognizer allows users to recognize ink content like shapes, handwriting, or the layout of inked documents targeting various scenarios like note-taking or document annotation.

- **Video Indexer**

  - This API can extract metadata from audio and video content automatically providing valuable insights like spoken words, faces, speakers, or even complete scenes.

- **Web search**

  - **Bing Autosuggest**

    - Bing Autosuggest offers intelligent type-ahead capabilities which can help end users to complete queries much faster and more efficiently.

  - **Bing Custom Search**

    - This service can be used to create a customized search engine which allows you to offer ad-free, commercial-grade results based on your scenarios.

  - **Bing Entity Search**

    - An API which offers the ability to recognize and classify named entities to search and find comprehensive results based on those entities.

  - **Bing Image Search**

    - Targeting developers seeking to integrate image search capabilities into apps, this API allows you to add image search options to apps and websites.

  - **Bing News Search**

    - Bing News Search empowers developers to add a customizable news search engine to websites or apps offering the ability to filter by topic, local news, or metadata.

15

- **Bing Spell Check**

  - An API which can be used to assist end users in identifying and fixing spelling mistakes in real time.

- **Bing Video Search**

  - Like the Bing Image Search, this API allows developers to add a range of advanced video search capabilities, like video previews, trending videos, or videos based on specific metadata to apps.

- **Bing Visual Search**

  - This service allows end users to search for content using images instead of regular text-based search queries.

- **Bing Web Search**

  - With Bing Web Search, developers can build solutions to retrieve documents, web pages, videos, images, or news stored on the Internet with a single API call.

As you can see, the list of Cognitive Services is quite long, and there are many different services which can be used in a variety of use cases. As this book is about the conversational aspect of artificial intelligence, we certainly will not cover all those Cognitive Services in detail. But some of those APIs will be explained more precisely in a later chapter, as many of them play an important role to build a conversational AI application.

## Solution Accelerators and Templates

When starting building bots using the Microsoft conversational AI platform, it is probably a good approach to use some kind of template instead of building everything from scratch. Microsoft did a good job in offering a variety of templates for developers and chatbot architects which can be used as a starting point for building a new project. Currently, the following templates are available for .NET, JavaScript/TypeScript, and Python development which are described in Table 1-1.

**Table 1-1.** *Bot Framework templates (source: https://github.com/microsoft/ BotBuilder-Samples/tree/master/generators)*

| Template | Description |
|---|---|
| Echo bot | A good template if you want a little more than "Hello World!" but not much more. This template handles the very basics of sending messages to a bot, and having the bot process the messages by repeating them back to the user. This template produces a bot that simply "echoes" back to the user anything the user says to the bot. |
| Core bot | The most advanced template, the core bot template provides six core features every bot is likely to have. This template covers the core features of a conversational AI bot using LUIS. |
| Empty bot | A good template if you are familiar with Bot Framework v4, and simply want a basic skeleton project. Also, it is a good option if you want to take sample code from the documentation and paste it into a minimal bot in order to learn. |

**Note**    All the preceding templates can be used in a variety of scenarios. If you are a .NET developer, you can use the Visual Studio extension from https:// github.com/Microsoft/BotBuilder-Samples/tree/master/ generators/vsix-vs-win/BotBuilderVSIX-V4 which installs the solution templates to your Visual Studio IDE.

With the extension installed, you can simply start a new Visual Studio Bot Framework project as with any other template, which is shown in Figure 1-4.

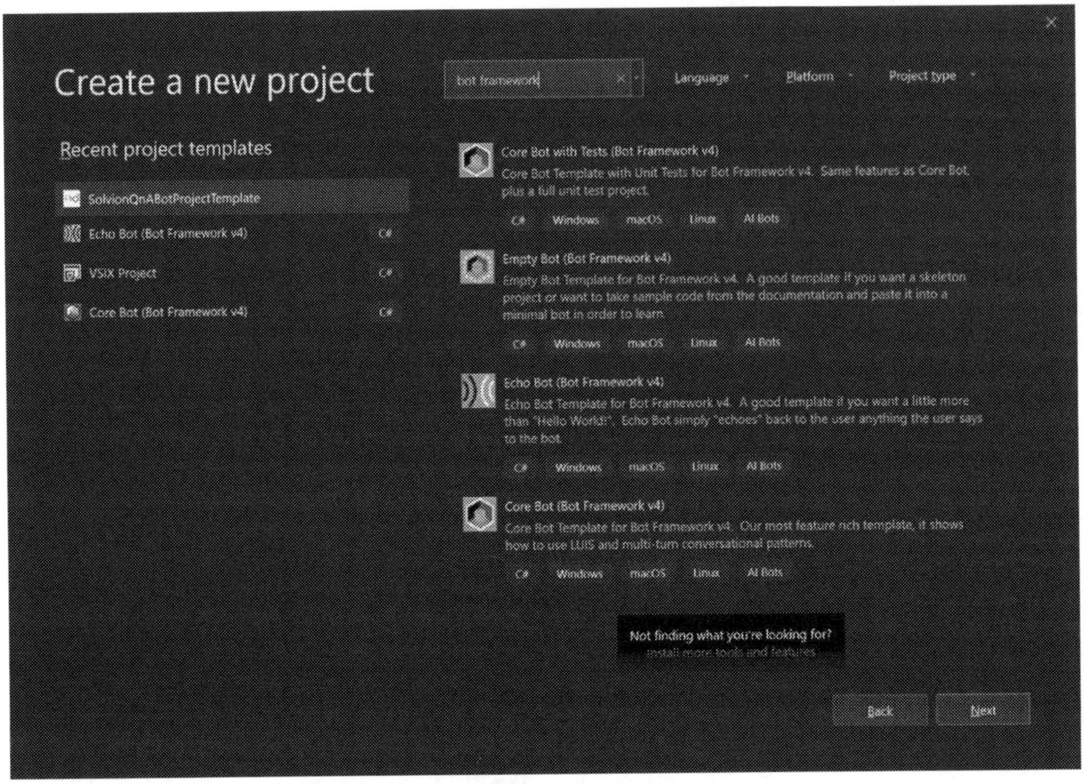

*Figure 1-4.  Bot Framework Visual Studio templates*

If you want to build bots using .NET core without using Visual Studio but any other web development IDE, you can utilize the .NET Core templates.

---

**Note**    The .Net Core Visual Studio Bot Framework templates can be installed from here: `https://github.com/microsoft/BotBuilder-Samples/tree/master/generators/dotnet-templates`.

---

These templates allow you to scaffold your bot project completely via the command line with the `dotnet new` command like this:

```
dotnet new echobot -n MyEchoBot
```

But if you are a web developer who is comfortable in developing solutions in JavaScript or TypeScript using Visual Studio Code, for example, you can use the Yeoman generator for Bot Framework v4 from here `https://github.com/microsoft/BotBuilder-Samples/tree/master/generators/generator-botbuilder` to scaffold a project using the CLI by simply executing a command like the following from your favorite command-line interface:

yo botbuilder

This command will start the Yeoman generator which will guide you through the setup and scaffolding process of your project shown in Figure 1-5.

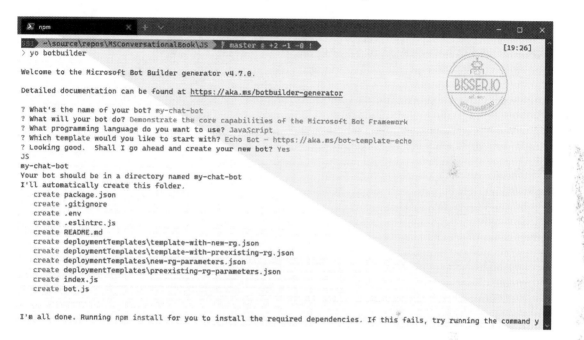

***Figure 1-5.*** *Bot Framework Yeoman generator for JavaScript/TypeScript*

A similar generator is also available for Python-based development, which can be installed from here https://github.com/microsoft/BotBuilder-Samples/tree/master/generators/python. This generator allows you to scaffold Python-based Bot Framework v4 bots using the command line which makes your development phase much more comfortable.

---

**Note**    All these templates provide you with a scaffolded project which can be further enhanced with the bot's logic you want to implement. In addition to those templates, Microsoft also offers a set of solution accelerators, which are basically ready-to-go bot projects. All of the following solutions can be found here: https://microsoft.github.io/botframework-solutions/index.

---

Table 1-2 lists all solution accelerators which are currently available through the Bot Framework SDK.

*Table 1-2.* *Bot Framework solution accelerators (source: $https://microsoft.$* *$github.io/botframework-solutions/index$)*

| Name | Description |
|---|---|
| **Virtual assistant** | At its core the virtual assistant (available in C# and TypeScript) is a project template with the best practices for developing a bot on the Microsoft Azure platform. |
| **Enterprise assistant** | The enterprise assistant sample is an example of a virtual assistant that helps conceptualize and demonstrate how an assistant could be used in common enterprise scenarios. It also provides a starting point for those interested in creating an assistant customized for this scenario. |
| **Hospitality assistant** | The hospitality sample builds off of the virtual assistant template with the addition of a QnA Maker knowledge base for answering common hotel FAQs and customized Adaptive Cards. |

The virtual assistant is the most advanced of the three templates which basically integrates everything a typical enterprise-grade bot offers. It is based on the Bot Framework skills, which allow you to integrate reusable skills in your bot as well as leverage already predefined skills, like an email skill, a calendar skill, a to-do skill, or a point of interest skill. This template is a ready-to-use bot which besides various skills also integrates different input and output types, like buttons, Adaptive Cards, and even speaking capabilities. This enables the virtual assistant to be deployed on various channels at the same time as well as many different devices, ranging from browsers and apps to smart devices and even vehicles, like cars and others. One of the major advantages of using the virtual assistant is the fact that you will have a sophisticated bot, containing a lot of business logic, in a matter of minutes up and running. This enables developers to streamline the development process quite extensively so that more time can be used to focus on the business requirements instead of spending a lot of time setting up the core environment. Figure 1-6 shows a complete picture of the virtual assistant solution accelerator and all used services attached to it.

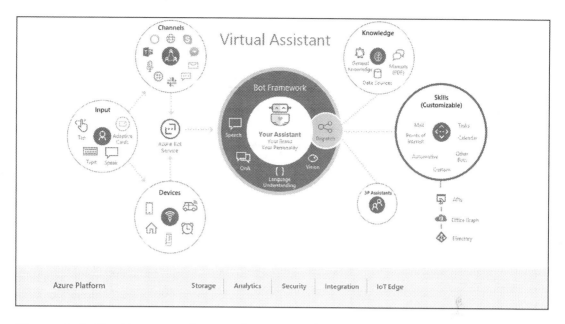

***Figure 1-6.*** *Bot Framework virtual assistant (source: https://microsoft.github.io/botframework-solutions/overview/virtual-assistant-solution/)*

As the virtual assistant is a quite holistic example of a bot, the enterprise assistant is an example of how an assistant can be used in an enterprise scenario. This template incorporates some of the most used skills and use cases like email and ITSM in an enterprise-grade bot project. The ITSM skill is based on the ServiceNow platform, which requires you to populate only a few configuration objects in the appsettings.json file of your bot to establish the connection between your bot and ServiceNow, and within a matter of minutes, your bot will be able to create new tickets, update existing tickets, or show the details of specific tickets within your helpdesk service, which is represented in Figure 1-7.

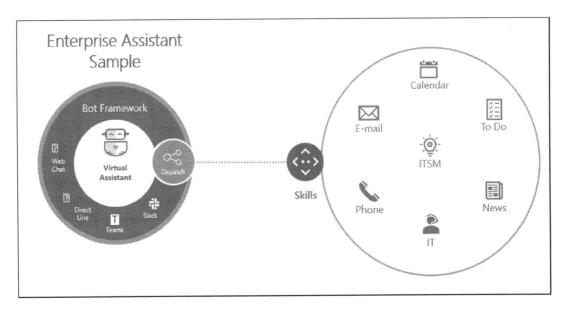

***Figure 1-7.***  *Bot Framework enterprise assistant (source:* `https://microsoft.`
`github.io/botframework-solutions/solution-accelerators/assistants/`
`enterprise-assistant)`

In contrast to the virtual assistant or enterprise assistant template, which is not
targeting a specific business vertical, the hospitality assistant can be used as a starting
point for building a bot specially for hospitality purposes like hotels or restaurants.
This template consists of a variety of skills like restaurant booking, events, news, or
weather skills, which are designed to be used in such a scenario. All the following skills,
which can be seen in the following illustration, are already ready to be used and can
be adapted to the specific requirements. For example, the weather skill is leveraging
the AccuWeather API to gather the weather information for a given location. All you as
a developer need to do is to provide your own API key into the appsettings.json file of
your bot, and the skill is fully functional, without the need of writing code. Figure 1-8
illustrates the skills attached to this solution accelerator sample.

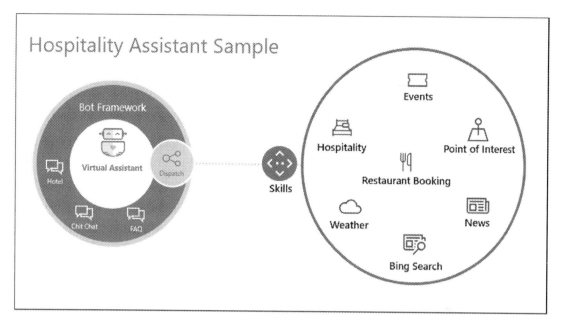

*Figure 1-8.   Bot Framework hospitality assistant (source: https://microsoft. github.io/botframework-solutions/solution-accelerators/assistants/ hospitality-assistant)*

## SaaS Solutions

The very top of the Microsoft conversational AI platform consists of AI-based Software as a Service solutions. A lot of these solutions can be found in Microsoft Teams for instance. If you take a look at the Teams app store, you will notice that there are a lot of bots available like WhoBot, a bot which connects you with people within your organization, and Polly, an AI-based assistant helping you to create and manage polls in a team. All these solutions are designed to assist you in different scenarios, without the need of developing or managing the bot. Many, if not all those bots you find in the Teams app store, are built on top of the Microsoft conversational AI platform, leveraging the Microsoft Bot Framework and Cognitive Services, to build a bot.

Another SaaS solution offered by Microsoft called Power Virtual Agents can help you to build bots. The difference between this solution and the solutions described in the preceding text is that with Power Virtual Agents, you do not need to write code. It offers a visual designer which lets you build dialog trees and conversational scenarios via a drag-and-drop interface, which generates a bot for you. This then can be used in a variety of scenarios like customer service to enhance your processes using prebuilt AI services.

# Summary

This chapter was a brief introduction to the Microsoft conversational AI platform. The first part of this chapter outlined the key parts of conversational AI applications focusing on natural language processing, translation, and speech. Then we took a look at the various usage scenarios for conversational AI and the benefits which come with that. The last part described the different parts and services within the conversational AI platform offered by Microsoft, like the Bot Framework or Cognitive Services, which will be covered in detail in the following chapters.

In the next chapter, we will take a more detailed look on the Microsoft Bot Framework, a platform designed to build and host comprehensive bots. There you will learn the key concepts of the framework as well as some best practices and tools to use when building bots using the Microsoft conversational AI platform.

# CHAPTER 2

# Introduction to the Microsoft Bot Framework

In March 2016, Microsoft announced the public preview of a project, originating from FUSE Labs, a group within Microsoft focusing on real-time and media-rich experiences, called the Microsoft Bot Framework. The main focus of this product was to provide what bot developers needed to build intelligent and interactive bots. Soon after that, the Bot Framework SDK version 3 has been made generally available, and a lot of developers started using this framework to build comprehensive bots. Almost two and a half years later, Microsoft released version 4 of its Bot Framework in late 2018 making it even easier to build bots across a variety of supported programming languages and platforms. With v4, the Bot Framework currently supports building bots using .NET, JavaScript/TypeScript, and Python as well as Java, which is still in preview.

By now, there are more than 50,000 active bots built with the Microsoft Bot Framework per month and more than 1.25 billion messages sent/received by those bots every month. With an availability of more than 99.9%, the Microsoft Bot Framework is one of the most used bot building platforms in the market, gaining popularity as we speak.

Therefore, this chapter should outline the core principles of this comprehensive framework and all available SDKs, tools, and services for building sophisticated bots.

## Key Concepts of the Microsoft Bot Framework

In general, a bot can be seen as a web application with some special integrations, as it is not designed to offer users only a graphical user interface but establish a humanlike conversation with end users. These conversations can be of different nature, including text, graphics like images or cards, and even speech. The core of these conversations are so-called activities, which represent on single interaction between the bot and the user. As the users typically interact with the bot through a channel, which basically is the application the bot is connected to, the Bot Framework Service is responsible for routing the activities between the bot and the channel. This Bot Framework Service is a component within the Azure Bot Service, handling those routing and messaging routines for you, so you do not need to take care of these while developing the bot.

When looking at the echo bot template, described in the previous chapter, the activity flow of such a conversation between the bot and the user may look as illustrated in Figure 2-1.

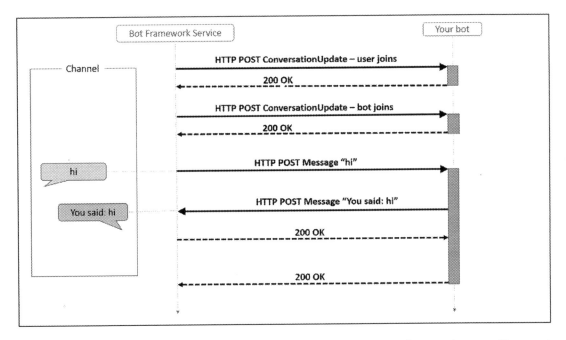

**Figure 2-1.** *Bot Framework activity flow (source: https://docs.microsoft.com/ en-us/azure/bot-service/bot-builder-basics)*

This simple conversation consists of two messages, one from the user to the bot say "hi" and one from the bot to the user echoing back "You said: hi." But there are more activities bound to that conversation. First, the Bot Framework Service sends an activity of type *ConversationUpdate* to the bot, indicating that a new user joined the conversation. After the bot sends an HTTP 200 OK back to the Bot Framework Service, acknowledging that the conversation can start, there is a second *ConversationUpdate* activity exchanged between the Bot Framework Service and the bot indicating that the bot is also joining the conversation. After that, the bot can receive the user's message. Now message themselves are also activities of type *Message*, which are exchanged between the channel and the bot routed via the Bot Framework Service. When the bot sends a message activity, the channel used by the user is responsible for rendering the message, which could include text, images, cards, or other rich media attachments like audio, video, buttons, or text to be spoken. In this simple example, the bot reacts on message activities sent by the user, but the bot can also respond to other activity types, like ConversationUpdate activities as well to greet the user upon entering the conversation, which is then called "proactive messaging," described in a later chapter.

As the preceding illustration outlines, activities are sent using HTTP POST requests. Although the protocol does not define the exact order of HTTP requests, the requests are nested. This means that outbound requests, meaning requests from the bot to the user, are executed within the scope of the inbound requests, meaning from the user to the bot. This makes sure that the order is kept while exchanging activities back and forth.

The Bot Framework SDK uses a pattern called *"turn"* to group an incoming activity with all outgoing activities belonging to that incoming activity. This can be compared to how people speak, which mainly is one at a time, speaking in turns. Therefore, the turn can be defined as the processing of a specific activity. Within a turn, there is a turn context object, which basically is an information store for activities. This turn context usually stores information about the sender, receiver, channel, and other necessary metadata, which is needed to process the activity successfully. The turn context is also accessible across all different layers of the bot, like the middleware components or the bot's logic, allowing these components to retrieve the activity details anytime. In addition to that, the turn context also allows the middleware components and the application logic of the bot to send outbound activities as well, making it one of the key components within the Bot Framework SDK.

# Activity Processing

Looking at Figure 2-2 which basically is showing the processing of the activity in an echo bot example mentioned in the preceding text, you'll see that there is a bit more that happens under the curtains. After the Bot Framework Service receives an inbound activity, it calls the responsible adapter's ProcessActivity method. Each activity which is send between the channel and the bot as an HTTP POST request always consists of a JSON payload. This payload is then deserialized as an activity object before it is passed to the adapter's ProcessActivity method. After the adapter has received the deserialized activity object, a new turn context is created. With that new turn context object, the adapter then calls the middleware components to handle that turn context. As the turn context offers the possibility to send outbound activities, it provides send, update, and delete response functions which run asynchronously.

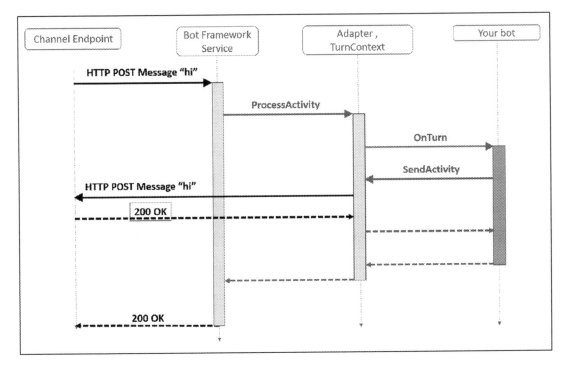

***Figure 2-2.*** *Bot Framework activity stack (source: `https://docs.microsoft. com/en-us/azure/bot-service/bot-builder-basics`)*

## Activity Handlers

When the bot then receives the new activity it receives, it calls the activity handlers. The main activity in the Bot Framework SDK is called the turn handler, which is the one component where all activities are getting routed through. The turn handler then calls all the other handlers, to handle the specific activity types, which are described in Table 2-1 for C#.

*Table 2-1.* *Bot Framework activity handlers C# (source: `https://docs.`*
*`microsoft.com/en-us/azure/bot-service/bot-builder-basics`)*

| Handler Name | Event | Description |
| --- | --- | --- |
| OnTurnAsync | Any activity type received | Calls one of the other handlers, based on the type of activity received. |
| OnMessageActivityAsync | Message activity received | Handles message activities. |
| OnConversationUpdate ActivityAsync | Conversation update activity received | On a conversationUpdate activity, calls a handler if members other than the bot joined or left the conversation. |
| OnMembersAddedAsync | Nonbot members joined the conversation | Handles members joining a conversation. |
| OnMembersRemovedAsync | Nonbot members left the conversation | Handles members leaving a conversation. |
| OnEventActivityAsync | Event activity received | On an event activity, calls a handler specific to the event type. |
| OnTokenResponseEventAsync | Token-response event activity received | Handles token-response events. |
| OnEventAsync | Non-token-response event activity received | Handles other types of events. |
| OnMessageReaction ActivityAsync | Message reaction activity received | On a messageReaction activity, calls a handler if one or more reactions were added or removed from a message. |
| OnReactionsAddedAsync | Message reactions added to a message | Handles reactions added to a message. |
| OnReactionsRemovedAsync | Message reactions removed from a message | Handles reactions removed from a message. |
| OnUnrecognizedActivity TypeAsync | Other activity type received | Handles any activity type otherwise unhandled. |

Table 2-2 outlines the different activity handlers for the Microsoft Bot Framework JavaScript SDK.

***Table 2-2.*** *Bot Framework activity handlers JavaScript (source: https://docs.microsoft.com/en-us/azure/bot-service/bot-builder-basics)*

| Handler Name | Event | Description |
| --- | --- | --- |
| **onTurn** | Any activity type received | Registers a listener for when any activity is received. |
| **onMessage** | Message activity received | Registers a listener for when a message activity is received. |
| **onConversationUpdate** | Conversation update activity received | Registers a listener for when any conversationUpdate activity is received. |
| **onMembersAdded** | Members joined the conversation | Registers a listener for when members joined the conversation, including the bot. |
| **onMembersRemoved** | Members left the conversation | Registers a listener for when members left the conversation, including the bot. |
| **onMessageReaction** | Message reaction activity received | Registers a listener for when any messageReaction activity is received. |
| **onReactionsAdded** | Message reactions added to a message | Registers a listener for when reactions are added to a message. |
| **onReactionsRemoved** | Message reactions removed from a message | Registers a listener for when reactions are removed from a message. |
| **onEvent** | Event activity received | Registers a listener for when any event activity is received. |
| **onTokenResponseEvent** | Token-response event activity received | Registers a listener for when a token-response event is received. |
| **onUnrecognizedActivityType** | Other activity type received | Registers a listener for when a handler for the specific type of activity is not defined. |
| **onDialog** | Activity handlers have completed | Called after any applicable handlers have completed. |

# State

From a bot perspective, the state can be seen as the mind or memory of the bot. With that in mind, the bot is able to remember things about the user or the conversation. This allows the bot to access specific information, which, for example, the user once wrote so that the bot does not need to ask for that specific information in a later conversation again. The information stored in the state is also typically longer available as the information within a given turn. This component therefore enables the bot to accomplish what's called multiturn conversations, meaning conversations which are more than one turn long. There are multiple different layers involved in the bot's state:

- **Storage layer**

- **State management layer**

- **State property accessors**

All those components are connected offering a sophisticated state management built into the Bot Framework SDK. Figure 2-3 should outline the state processing sequence and the various components interacting with each other.

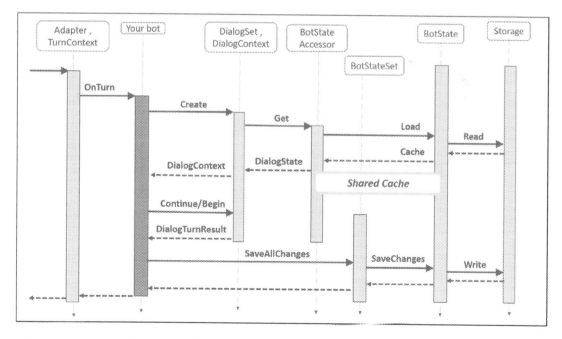

***Figure 2-3.*** *Bot Framework state management (source:* `https://docs.` `microsoft.com/en-us/azure/bot-service/bot-builder-concept-state`)

The solid arrows in the preceding figure indicate individual method calls executed by specific bot components, whereas the dashed arrows indicate the resulting responses. The state is usually stored in the storage layer, which is the rightest component in this figure. There are many ways of how to integrate a storage solution for storing the state like in-memory, databases, or other file storage solutions. The Bot Framework SDK already has some of these solutions built-in:

- **Memory storage**

  - The memory storage should be used for testing purposes, as it is designed and intended for testing the bot locally as this type of storage is quite temporary. Another downside of this is that when you are running your bot in Azure as a web app, you typically pay for the amount of resources you allocate. This could lead to higher costs, if you are planning to use memory for storing the state as you would need to allocate quite a high amount of memory. Another disadvantage here is that the state will be deleted every time the bot or the web app containing the code is restarted due to updates or service outages.

- **Azure Blog storage**

  - The Azure Blob storage functionality offers the possibility to store the state information in blobs within an Azure Storage Account. This has the advantage that you can persist the data and separate the bot's logic from the state storage components.

- **Azure Cosmos DB storage**

  - In larger scenarios, it makes sense to use an Azure Cosmos DB instead of an Azure Blob storage to store the state information in, as the Cosmos DB typically offers more flexibility in terms of scaling and availability, but this of course also can lead to higher costs compared to Azure Blog storage.

The state management component within the SDK provides automated reading and writing of the state within your bot connected to the storage layer you defined. The state information itself is stored as state properties. These state properties are key-value pairs, which allows you to define the object structure as you can define state properties as classes. Consequently, when retrieving state properties, you know how the structure of that object data makes it easier for you to handle them within your code. The state properties are out of the box categorized into three different groups:

- **User state**

  - This type of state property can be used to store all information about a user communicating with the bot in that specific channel. The user state is accessible in every turn with no dependency to the current conversation. Therefore, this state property should be used to store user information like the name, preferences, or settings specific to the user as well as information about previous conversations the user had with the bot. This can then be persisted throughout the lifetime of a single conversation.

- **Conversation state**

  - The conversation state property is available in any turn within a specific conversation regardless of the user. This means it can be used in group conversation scenarios as well (like conversations in Microsoft Teams channels). Therefore, this property should be used to store information like what questions the bot already asked the user or what the topic of the current conversation is.

- **Private conversation state**

  - This property can be seen as a combination of the preceding two, meaning it is targeting a specific conversation and a specific user. Accordingly, it is designed for group conversations especially as you can store user-specific conversation information in group chats within that property.

To access the state properties, the Bot Framework SDK offers so-called state property accessors. These components are designed to enable reading and writing of state properties within your bot offering *get, set, and delete* methods within turns. If you use the get method within a turn to get a specific state property, the SDK will load that information into the local bot cache. Hence, it is stored within the local cache because it is more performant, than reaching out to the storage every time to retrieve state information. Additionally, the accessors allow you to treat the state properties like local variables, meaning you can manipulate them within your code. In order to persist the locally cached state property, you have to use the *SaveChanges* method provided by the accessor to save the information stored in the locally cached state object to the storage. It is important to remember that only those properties will be persisted in the storage, which have been set up within the state group without affecting the others. Thus, it is possible to store all the information from the local cache to the conversation state in the storage, without storing any information to the user state in the storage. To avoid conflicts, the last write wins, meaning that the write transaction with the last write timestamp will overwrite any other transaction made earlier for a given property.

# Dialogs

Dialogs are basically the core of your bot, as they are used to manage the conversation and guide the user through the conversation tree. A dialog can be compared to functions within your bot which basically process specific tasks in a given sequence. The trigger of dialogs can vary, sometimes they are executed based on a specific user message, other times the trigger is another service indicating that a new part of the conversation shall be started, or one dialog calls another one to continue the conversation. As dialogs are one of the key notions within the Bot Framework SDK, there are already some prebuilt functions like prompts or waterfall-like dialogs included, which can be used with little implementation efforts. When building a bot using the Bot Framework, one good approach would be to sketch out the dialog tree which should be implemented to gain an idea which parts of the conversation you need to implement in your code. Figure 2-4 outlines the supported dialog types and prompt features offered by the SDK natively.

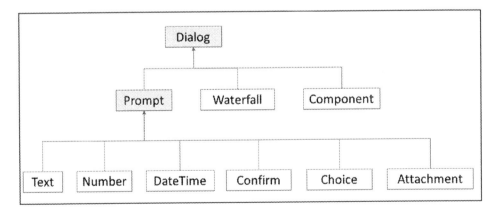

*Figure 2-4. Bot Framework prebuilt dialog and prompt types (source: https:// docs.microsoft.com/en-us/azure/bot-service/bot-builder-concept-dialog)*

Prompts are the simplest form of collecting information from the user. The prompts' functionality offers many different prompt types, which are outlined in Table 2-3.

***Table 2-3.*** *Bot Framework prompt types (source: https://docs.microsoft.com/ en-us/azure/bot-service/bot-builder-concept-dialog)*

| Prompt | Description | Returns |
| --- | --- | --- |
| **Attachment prompt** | Asks for one or more attachments, such as a document or image. | A collection of attachment objects. |
| **Choice prompt** | Asks for a choice from a set of options. | A found choice object. |
| **Confirm prompt** | Asks for a confirmation. | A Boolean value. |
| **Date-time prompt** | Asks for a date-time. | A collection of date-time resolution objects. |
| **Number prompt** | Asks for a number. | A numeric value. |
| **Text prompt** | Asks for general text input. | A string. |

A prompt is the most basic form of a dialog as it has only two steps. First, the prompt is triggered asking the user for input. Second, the input is evaluated, and the response is being returned. Each prompt has prompt options which can be used to define the prompt text, the retry prompt (this is the prompt which will be triggered if the validation is not successful), and if it is a choice prompt, the answer choices are included as well in the options. If you need to include your own validation rules in addition to the predefined ones, you can do so by simply adding those custom validators to your prompt. The prompt will then execute the predefined validator first and then the custom validator. One simple example would be to prompt the user for a flight destination. Now in that case, you would most probably use the text prompt to ask for an airport name. But the validation here could be quite tricky, as the default text validator would only check if the input is a text. But this means that users could also insert any kind of text, which is not really what you want. Therefore, you could implement a custom validator to check the input against a collection of airport names and abbreviations to see if the user has entered a valid airport name before proceeding the conversation. If the validation fails, the prompt will call the retry prompt to ask the user again for an airport name. Choice, confirm, date-time, and number prompts also use the prompt locale to discover language-specific circumstances. The locale is retrieved from the channel the bot is connected to, allowing to enable multilanguage scenarios within prompts.

Waterfall dialogs are mainly used to gather information in a waterfall-like manner. The goal of a waterfall dialog is to ask the user a set of questions in the form of prompts, in order to collect the needed information. The key point here is that the bot does not ask one question to get a collection of responses, but to ask for each response separately in a series of questions. After the bot prompts the user with a question, the user needs to respond to that question. If the bot has received and successfully validated that response, the next prompt will be executed. This is done until all the prompts have been answered, and the needed information has been gathered, which could look like the one Figure 2-5 is indicating.

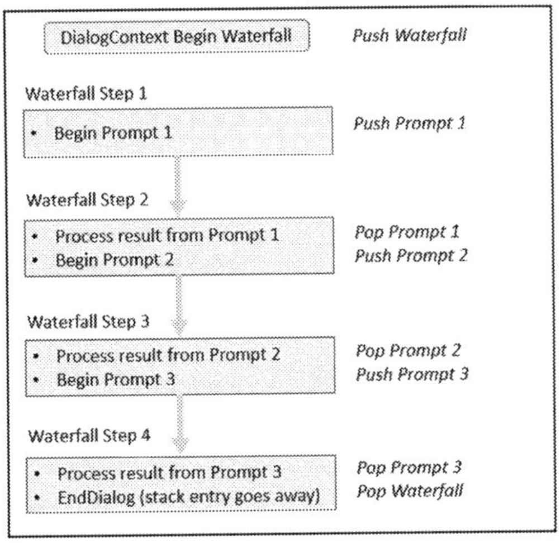

*Figure 2-5.* *Bot Framework waterfall dialog example (source:* `https://docs.microsoft.com/en-us/azure/bot-service/bot-builder-concept-dialog`*)*

Within a single waterfall dialog, there is a so-called waterfall step context used to store and access the turn context as well as the state. As the waterfall dialog, basically, consists of a series of prompts, you should also think about the validation of user responses for specific prompts in order to avoid that the next step is being called with an invalid value. This can be done using the prompt validator context parameter, which is part of the prompt functionality. This parameter returns a Boolean value indicating if the user's response validation has been successful or not.

The waterfall dialogs are somewhat designed to serve for a specific use case. But the SDK also offers a dialog type called the *component dialog* which is designed to be reusable. This type of dialog is intended to be somewhat independent which allows you to create a component dialog for a given use case, which you can then export as a package, for example, to include it in other bots as well. Within a component dialog, you can include a set of waterfall dialogs or prompts which are grouped together as a dialog set.

The component which is responsible for managing dialogs is called the dialog context. This dialog context offers methods to begin, replace, continue, end, or cancel a dialog. You can think of all the dialogs within your bot as a dialog stack, whereas the turn handler mentioned in the preceding text is the component controlling the dialog stack. If the dialog stack is empty at some point, the turn handler also acts as a fallback to continue the conversation.

Upon beginning a new dialog, it is being pushed to the top of the dialog stack and will become the active dialog. It remains active until the dialog either ends, is canceled, or is removed from the stack as it has been replaced by another dialog within the stack. The replacement of a dialog usually happens via the replace dialog method within the active dialog. After the dialog which replaced the previous dialog ends, it is being removed from the dialog stack, and the previous dialog becomes the active dialog again as it is again the top of the stack. This enables you to branch dialogs in a given conversation, which is helpful in scenarios like user interruption or similar.

# Middleware

The middleware component is implemented between the adapter and the bot. The middleware concept has been designed to manipulate activities either before or after a given turn. Therefore, the middleware components can be seen as pre- or postactivity processing engines within your bot's activity pipeline. During the initialization of your bot, after starting or restarting it, the adapter adds all configured middleware components to its middleware collection. The order in which the adapter adds the middleware components to its stack defines which middleware component will be called in which order. Figure 2-6 outlines the general middleware processing concept.

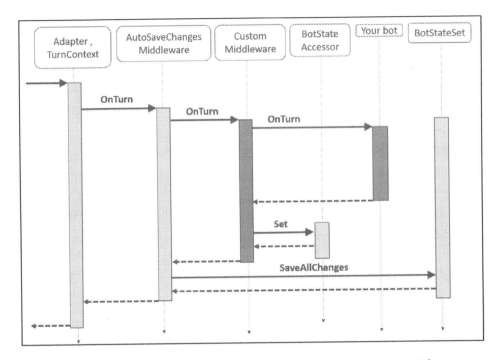

***Figure 2-6.*** *Bot Framework middleware concept (source:* `https://docs.`
`microsoft.com/en-us/azure/bot-service/bot-builder-concept-middleware`*)*

If you implement multiple middleware components within your project, the order
in which the middleware components are called is quite important. You could, for
example, implement a middleware which should log all information connected to an
activity into an Azure storage or Application Insights. And then you would like to use
another middleware component to translate a message activity which is not received
in English, as your bot and the knowledge base it is connected to only support English
as a language. Therefore, you may want to log all the information which is received by
the user in the original language to the Azure service before it is translated to keep the
original messages stored. Moreover, middleware components are used to manipulate
not only incoming activities but also outgoing ones. Sticking with the preceding example,
we could use the same middleware responsible to translate incoming messages and use
it to translate outgoing messages as well to fully support multilingual scenarios.

As the preceding illustration shows, each middleware implements a call *next()*
indicating to execute the next layer, which can either be another middleware or the bot
logic. If this next method is not called, the following layers are not executed, which is
called short circuiting. This can be beneficial in scenarios where you want to cancel
the current turn before it is handled by your turn handler, due to a failure or similar.

The result of this is that, although the turn handler is not called, the logic within the middleware which is executing the short circuiting is still executed and the bot's conversation is in a safe state.

# Bot Project Structure

The following paragraphs will walk you through the basic project structure of a simple echo bot in C# and JavaScript. More detailed explanations and enhancements will be discussed in a later chapter.

## Echo Bot Logic C#

After creating your Bot Framework echo bot project, using either the Visual Studio Bot Framework extensions mentioned in the preceding text or the .NET Core CLI, the typical project structure of your core bot will look like in Figure 2-7.

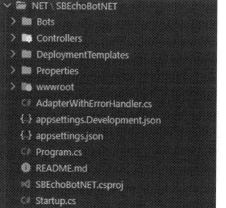

*Figure 2-7.* *Bot Framework C# Echo Bot Project Structure*

---

**Note**   For a detailed list of prerequisites for developing a bot using the Bot Framework SDK for .NET, please conduct `https://docs.microsoft.com/en-us/azure/bot-service/dotnet/bot-builder-dotnet-sdk-quickstart`.

---

The *Program.cs* is basically the entry point of the bot project, creating a new defining the host to be used as well as the startup class to be used, which in our case is the *Startup.cs*:

```
using Microsoft.AspNetCore;
using Microsoft.AspNetCore.Hosting;

namespace SBEchoBotNET
{
    public class Program
    {
        public static void Main(string[] args)
        {
            CreateWebHostBuilder(args).Build().Run();
        }

        public static IWebHostBuilder CreateWebHostBuilder(string[] args) =>
            WebHost.CreateDefaultBuilder(args)
                .UseStartup<Startup>();
    }
}
```

The *Startup.cs* is the class which defines all configuration components. Additionally, it also creates the bot as a transient for us to use it within our project:

```
using Microsoft.AspNetCore.Builder;
using Microsoft.AspNetCore.Hosting;
using Microsoft.AspNetCore.Mvc;
using Microsoft.Bot.Builder;
using Microsoft.Bot.Builder.Integration.AspNet.Core;
using Microsoft.Bot.Connector.Authentication;
using Microsoft.Bot.Builder.BotFramework;
using Microsoft.Extensions.Configuration;
using Microsoft.Extensions.DependencyInjection;
using SBEchoBotNET.Bots;
```

```
namespace SBEchoBotNET
{
    public class Startup
    {
        public Startup(IConfiguration configuration)
        {
            Configuration = configuration;
        }

        public IConfiguration Configuration { get; }

        // This method gets called by the runtime. Use this method to add
        // services to the container.
        public void ConfigureServices(IServiceCollection services)
        {
            services.AddMvc().SetCompatibilityVersion(CompatibilityVersion.
            Version_2_1);
            // Create the Bot Framework Adapter with error handling enabled.
            services.AddSingleton<IBotFrameworkHttpAdapter,
            AdapterWithErrorHandler>();
            // Create the bot as a transient. In this case the ASP
            // Controller is expecting an IBot.
            services.AddTransient<IBot, EchoBot>();
        }

        // This method gets called by the runtime. Use this method to
        // configure the HTTP request pipeline.
        public void Configure(IApplicationBuilder app, IHostingEnvironment env)
        {
            if (env.IsDevelopment())
            {
                app.UseDeveloperExceptionPage();
            }
            else
            {
                app.UseHsts();
            }
```

```
            app.UseDefaultFiles();
            app.UseStaticFiles();
            app.UseWebSockets();
            //app.UseHttpsRedirection();
            app.UseMvc();
        }
    }
}
```

As you can see in the preceding code, the line *services.AddSingleton<IBotFramew orkHttpAdapter, AdapterWithErrorHandler>();* adds the *AdapterWithErrorHandlers* to our services. The adapter is responsible for connecting the bot to a service endpoint, like the Azure Bot Service using the Bot Connector Service. Moreover, the adapter is used to cover authentication mechanisms, and it is used to exchange activities between the bot and the Bot Connector Service. After the adapter has received a new activity, it first creates the turn context and then calls the bot logic passing the created turn context for further processing. After the bot logic has processed the activity, the adapter sends the outbound activity back to the channel. Furthermore, the adapter is responsible for managing and calling the middleware pipeline to pre- or postprocess activities before or after a turn. As, in the echo bot template, there is no predefined middleware added, the *AdapterWithErrorHandlers.cs* looks as follows:

```
using Microsoft.Bot.Builder.Integration.AspNet.Core;
using Microsoft.Bot.Builder.TraceExtensions;
using Microsoft.Extensions.Configuration;
using Microsoft.Extensions.Logging;

namespace SBEchoBotNET
{
    public class AdapterWithErrorHandler : BotFrameworkHttpAdapter
    {
        public AdapterWithErrorHandler(IConfiguration configuration,
        ILogger<BotFrameworkHttpAdapter> logger)
            : base(configuration, logger)
```

```
    {
        OnTurnError = async (turnContext, exception) =>
        {
            // Log any leaked exception from the application.
            logger.LogError(exception, $"[OnTurnError] unhandled error :
            {exception.Message}");
            // Send a message to the user
            await turnContext.SendActivityAsync("The bot encountered an
            error or bug.");
            await turnContext.SendActivityAsync("To continue to run
            this bot, please fix the bot source code.");
            // Send a trace activity, which will be displayed in the
            Bot Framework Emulator
            await turnContext.TraceActivityAsync("OnTurnError Trace",
            exception.Message, "https://www.botframework.com/schemas/
            error", "TurnError");
        };
    }
  }
}
```

The bot logic usually sits in the *EchoBot.cs* file, which in our case is very slim. As the echo bot is only intended to echo back what the user says, we only have two methods in our bot logic:

- **OnMessageActivityAsync**

  - This method is one of the C# activity handlers mentioned previously used to handle messages within a given turn.

- **OnMembersAddedAsync**

  - This method is another C# activity handler mentioned before used to handle members joining the conversation.

The *OnMessageActivityAsync()* method will echo back the input from the user, whereas the *OnMembersAddedAsync()* method will greet the user with "Hello and welcome!" upon entering the conversation in a proactive manner. The complete implementation of our *EchoBot.cs* looks as follows:

```csharp
using System.Collections.Generic;
using System.Threading;
using System.Threading.Tasks;
using Microsoft.Bot.Builder;
using Microsoft.Bot.Schema;

namespace SBEchoBotNET.Bots
{
    public class EchoBot : ActivityHandler
    {
        protected override async Task OnMessageActivityAsync(ITurnContext
          <IMessageActivity> turnContext, CancellationToken
          cancellationToken)
        {
            var replyText = $"Echo: {turnContext.Activity.Text}";
            await turnContext.SendActivityAsync(MessageFactory.
            Text(replyText, replyText), cancellationToken);
        }

        protected override async Task OnMembersAddedAsync(IList<ChannelA
        ccount> membersAdded, ITurnContext<IConversationUpdateActivity>
        turnContext, CancellationToken cancellationToken)
        {
            var welcomeText = "Hello and welcome!";
            foreach (var member in membersAdded)
            {
                if (member.Id != turnContext.Activity.Recipient.Id)
                {
                    await turnContext.SendActivityAsync(MessageFactory.
                    Text(welcomeText, welcomeText), cancellationToken);
                }
            }
        }
    }
}
```

The last piece missing within our project is the *BotController.cs*. This controller defines the routing of messages as well as HTTP calls within our bot. As you can see in the following, the only route defined for our bot is *"api/messages."* The *PostAsync()* method calls the adapter to process the incoming HTTP requests. Therefore, when communicating with the bot, the channel or application you are using should send all HTTP requests to the API/messages route to make sure they are received by your bot:

```
using System.Threading.Tasks;
using Microsoft.AspNetCore.Mvc;
using Microsoft.Bot.Builder;
using Microsoft.Bot.Builder.Integration.AspNet.Core;

namespace SBEchoBotNET.Controllers
{
    // This ASP Controller is created to handle a request. Dependency
    Injection will provide the Adapter and IBot
    // implementation at runtime. Multiple different IBot implementations
    running at different endpoints can be
    // achieved by specifying a more specific type for the bot constructor
    argument.
    [Route("api/messages")]
    [ApiController]
    public class BotController : ControllerBase
    {
        private readonly IBotFrameworkHttpAdapter Adapter;
        private readonly IBot Bot;
        public BotController(IBotFrameworkHttpAdapter adapter, IBot bot)
        {
            Adapter = adapter;
            Bot = bot;
        }
        [HttpPost, HttpGet]
        public async Task PostAsync()
```

```
        {
            // Delegate the processing of the HTTP POST to the adapter.
            // The adapter will invoke the bot.
            await Adapter.ProcessAsync(Request, Response, Bot);
        }
    }
}
```

The configuration settings are stored in a file called *appsettings.json* usually. This file should hold all IDs, secrets, or connection strings used within your bot. The app settings of our echo bot will mainly look like this for now and will be adapted in a later chapter:

```
{
  "MicrosoftAppId": "",
  "MicrosoftAppPassword": ""
}
```

## Echo Bot Logic JavaScript

To create a Bot Framework project for developing a bot using JavaScript, you need to create your bot project using the Yeoman generator. Within your command line, navigate to the directory where you want to create the project and execute the following commands to install the prerequisites:

```
npm install -g npm
npm install -g yo
npm install -g generator-botbuilder
# only run this command if you are on Windows. Read the above note.
npm install -g windows-build-tools
```

> **Note**    For a detailed list of prerequisites for developing a bot using the Bot Framework SDK for JavaScript, please conduct https://docs.microsoft.com/en-us/azure/bot-service/javascript/bot-builder-javascript-quickstart.

After the prerequisites have been installed successfully, you need to run the following command and answer the questions asked by the Yeoman generator to create a new Bot Framework project:

```
yo botbuilder
```

If you are developing a JavaScript bot, the scaffolded project after running the Yeoman generator will look like shown in Figure 2-8.

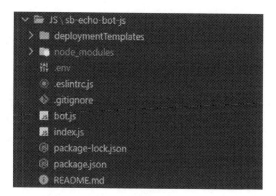

***Figure 2-8.***  *Bot Framework JS Echo Bot Project Structure*

You may notice that this is far more streamlined compared to the C# template, which lies in the nature of JavaScript as a programming language. What is separated into multiple files in C# is all stored in one file within your JS project. The index.js is using some dependencies to be included in the bot project which you can see here:

```
const dotenv = require('dotenv');
const path = require('path');
const restify = require('restify');
// Import required bot services.
const { BotFrameworkAdapter } = require('botbuilder');
// This bot's main dialog.
const { MyBot } = require('./bot');
// Import required bot configuration.
const ENV_FILE = path.join(__dirname, '.env');
dotenv.config({ path: ENV_FILE });
```

The *index.js* file holds the implementation for your adapter:

```
// Create adapter.
// See https://aka.ms/about-bot-adapter to learn more about how bots work.
const adapter = new BotFrameworkAdapter({
    appId: process.env.MicrosoftAppId,
    appPassword: process.env.MicrosoftAppPassword
});
```

The HTTP server is also created in that file using a JavaScript library called *restifiy*:

```
// Create HTTP server
const server = restify.createServer();
server.listen(process.env.port || process.env.PORT || 3978, () => {
    console.log(`\n${ server.name } listening to ${ server.url }`);
    console.log(`\nGet Bot Framework Emulator: https://aka.ms/botframework-
    emulator`);
    console.log(`\nTo test your bot, see: https://aka.ms/debug-with-
    emulator`);
});
```

Additionally, the *index.js* creates the bot controller as follows:

```
// Listen for incoming requests.
server.post('/api/messages', (req, res) => {
    adapter.processActivity(req, res, async (context) => {
        // Route to main dialog.
        await myBot.run(context);
    });
});
```

Furthermore, it instantiates the bot and its main dialog to be called each when a new activity arrives:

```
// Create the main dialog.
const myBot = new MyBot();
```

The second file which should be looked at is the *bot.js* file. This file also includes two methods similar to the C# implementation which are the following:

- **onMessage**

    - This method is one of the JavaScript activity handlers mentioned previously used to handle messages within a given turn.

- **onMembersAdded**

    - This method is another JavaScript activity handler mentioned before used to handle members joining the conversation.

In our case the *onMembersAdded()* function will greet every new user with "Hello and welcome!" upon entering the conversation, whereas the *onMessage()* function will echo back what the user said:

```javascript
const { ActivityHandler } = require('botbuilder');

class MyBot extends ActivityHandler {
    constructor() {
        super();
        // See https://aka.ms/about-bot-activity-message to learn more
        about the message and other activity types.
        this.onMessage(async (context, next) => {
            await context.sendActivity(`You said '${ context.activity.text
            }'`);
            // By calling next() you ensure that the next BotHandler is run.
            await next();
        });

        this.onMembersAdded(async (context, next) => {
            const membersAdded = context.activity.membersAdded;
            for (let cnt = 0; cnt < membersAdded.length; ++cnt) {
                if (membersAdded[cnt].id !== context.activity.recipient.id) {
                    await context.sendActivity('Hello and welcome!');
                }
            }
```

```
        // By calling next() you ensure that the next BotHandler is run.
        await next();
    });
  }
}
module.exports.MyBot = MyBot;
```

The file which is used to store the configuration is called the *.env* file in JavaScript projects, which can be compared to the appsettings.json in C# projects. This file usually stores all the configuration settings like the appId or the appPassword or other connection strings used within the bot. The echo bot's .env file will basically look like this after creating the project:

```
MicrosoftAppId=
MicrosoftAppPassword=
```

As this chapter should outline the basic principles of the bot components mentioned in the preceding text, it should indicate the key factors within a Bot Framework project. In later chapters, we will walk through some of these again, looking on how to adapt them to enhance the bot's functionality.

# Bot Framework Skills: Reusable Bot Components

Skills within the Bot Framework are reusable components for a bot including conversational items. The Bot Framework offers an extensibility model where developers can work on skills independently, which can then be integrated into a bot. This provides the flexibility of developing and maintaining an enterprise bot scenario which integrates different skills for different use cases. In the past you had two options to solve that. Either you could create one bot containing all the use case functionality in its logic or create separate bots for various use cases. The second option soon seemed to be quite uncomfortable for end users, as they would then need to remember which bot to use for which use case. By introducing the skill concept, developers could work on skills separately which then could be included into a bot using one single command-line operation which executes the dispatch and configuration changes.

A skill itself is also a bot, which could also be used standalone. The Microsoft Bot Framework SDK offers templates in C# and TypeScript for creating new skills using templates. Table 2-4 outlines the available skills offered in the SDK today, which can be used right away or adapted as needed.

***Table 2-4.*** *Predefined Bot Framework skills (source: `https://docs.microsoft.com/en-us/azure/bot-service/bot-builder-skills-overview`)*

| Name | Description |
|------|-------------|
| **Calendar Skill** | Add calendar capabilities to your assistant. Powered by Microsoft Graph and Google. |
| **Email Skill** | Add email capabilities to your assistant. Powered by Microsoft Graph and Google. |
| **To-Do Skill** | Add task management capabilities to your assistant. Powered by Microsoft Graph. |
| **Point of Interest Skill** | Find points of interest and directions. Powered by Azure Maps and FourSquare. |
| **Automotive Skill** | Industry-vertical skill for showcasing enabling car feature control. |
| **Experimental Skills** | News, restaurant booking, and weather. |

Throughout the next chapters, we will take a detailed look on how to develop and integrate those skills in a Bot Framework bot.

# Azure Bot Service: A Bot Hosting Platform

As mentioned in the previous chapter, the Azure Bot Service is the hosting platform for a bot developed using the Bot Framework. It is the glue, connecting your bot to the supported channels, establishing, and managing those connections for you. This enhances the process of deploying your bot to various channels, as, in some cases, the enablement of a channel is done within a few clicks and the bot is ready to be used in that channel. The current list of supported channels can be found in Chapter 1. But the Azure Bot Service is used not only to connect your bot to channels but also to manage the bot's configuration within the Azure portal (`https://portal.azure.com`).

It offers a complete management interface to manage everything related to your bot within that portal, like the bot handle or display name, the messaging endpoint, the Microsoft app ID, and app password used within your bot to secure the communication or analytics configuration settings used to define which Application Insights service to use to analyze the bot's performance and behavior, demonstrated in Figure 2-9.

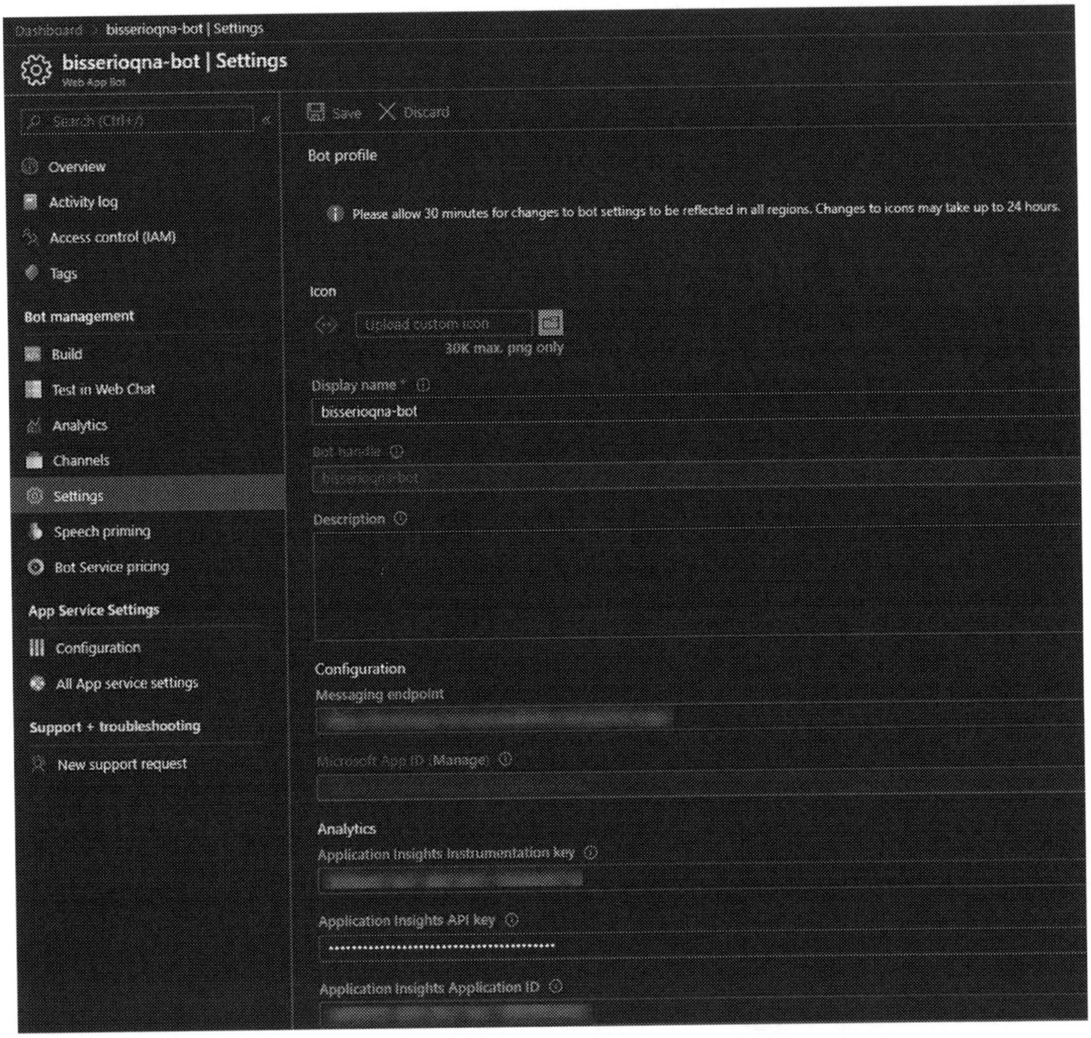

*Figure 2-9.* *Azure Bot Service management interface*

Therefore, the Azure Bot Service is the best option to integrate your bot within various channels and host the bot as a web application either in Azure, which is the preferred way or anywhere else, like in your on premises datacenter, or any other cloud service provider.

# Bot Framework SDK Tool Offerings

The following subchapters will outline the available tools within the Bot Framework, as we will use all of them in the later chapters, when looking on how to design, develop, or deploy a bot built with the Bot Framework SDK.

## Bot Framework Emulator

One of the tools offered within the Bot Framework ecosystem is the Bot Framework Emulator. This is a comprehensive bot testing and debugging desktop application targeting local bot development and testing. Additionally, with the emulator you can also test and debug bots hosted in Microsoft Azure or anywhere else, which gives you more details in terms of debugging a bot developed using the SDK. As the Bot Framework itself is defined to cover many different development platforms, the emulator is also available across many platforms:

- Windows

- OS X

- Linux

---

**Note**    The emulator can be downloaded from the respective GitHub repo, using this link `https://aka.ms/botemulator`.

---

Developers can use the emulator to test the bot running locally. Moreover, the developers can use the emulator to manage the services connected to the bot, like LUIS or QnA Maker, as well as trace and debug the activity stack as seen in Figure 2-10.

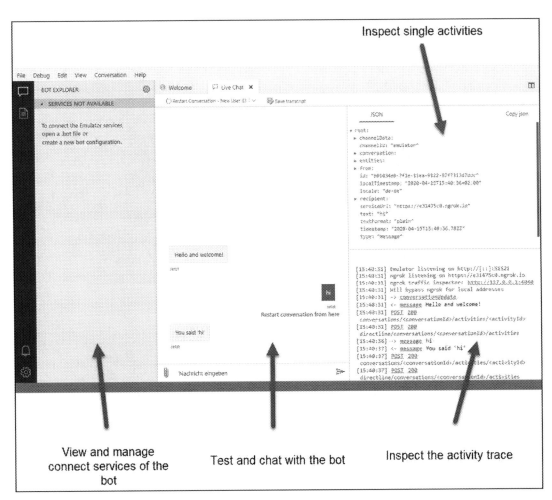

***Figure 2-10.*** *Bot Framework Emulator overview*

## Bot Framework Web Chat

As Bot Framework bots can be integrated into many different channels, there is also a web chat solution built by Microsoft available. The Bot Framework Web Chat component, which can be downloaded and used from here `http://aka.ms/bfwebchat`, is a highly customizable component built to render a web chat within an HTML page. Developers can either use the web chat directly as HTML/JavaScript component rwithin HTML code or use the React component built by Microsoft to even further customize the look and feel and behavior of the web chat client.

In order to integrate the web chat component using HTML and JavaScript, you typically would need to create the following lines to your website:

```
<!DOCTYPE html>
<html>
  <head>
    <script src="https://cdn.botframework.com/botframework-webchat/
    latest/webchat.js"></script>
    <style>
      html,
      body {
        height: 100%;
      }
      body {
        margin: 0;
      }

      #webchat {
        height: 100%;
        width: 100%;
      }
    </style>
  </head>
  <body>
    <div id="webchat" role="main"></div>
    <script>
      window.WebChat.renderWebChat(
        {
          directLine: window.WebChat.createDirectLine({
            token: 'YOUR_DIRECT_LINE_TOKEN'
          }),
          userID: 'YOUR_USER_ID',
          username: 'Web Chat User',
          locale: 'en-US',
          botAvatarInitials: 'WC',
          userAvatarInitials: 'WW'
        },
```

```
        document.getElementById('webchat')
    );
  </script>
  </body>
</html>
```

The preceding code would integrate the web chat component within your HTML page and connect it to your Bot Framework bot using the DirectLine token. The details of DirectLine protocol are explained in an upcoming chapter, but the DirectLine API basically establishes a secure communication between your bot and the application you integrate it into.

If you are looking into a more customizable approach in terms of styling and behavior, you could use the React component to integrate the web chat into your web application, using the following lines of code for instance:

```
import React, { useMemo } from 'react';
import ReactWebChat, { createDirectLine } from 'botframework-webchat';

export default () => {
  const directLine = useMemo(() => createDirectLine({ token: 'YOUR_DIRECT_
  LINE_TOKEN' }), []);

  return <ReactWebChat directLine={directLine} userID="YOUR_USER_ID" />;
};
```

To get an idea, which integration method offers which functionality, Table 2-5 outlines the key features for both scenarios.

**Table 2-5.**  *Bot Framework Web Chat feature comparison (source:* `https://github.com/Microsoft/BotFramework-WebChat`)

| Feature/Functionality | CDN Bundle | React |
|---|:---:|:---:|
| Change colors | ✓ | ✓ |
| Change sizes | ✓ | ✓ |
| Update/replace CSS styles | ✓ | ✓ |
| Listen to events | ✓ | ✓ |
| Interact with hosting webpage | ✓ | ✓ |
| Custom render activities | | ✓ |
| Custom render attachments | | ✓ |
| Add new UI components | | ✓ |
| Recompose the whole UI | | ✓ |

To give you an idea, the web chat, for example, could be customized to show custom user and bot avatar images as shown in Figure 2-11, enabling a more personal look and feel for users.

**Figure 2-11.**  *Bot Framework Web Chat customization example*

An end-to-end implementation guideline will be outlined in a later chapter, including all steps necessary to customize the web chat component.

# Bot Framework CLI

The Bot Framework Command-Line Interface is a cross-platform CLI, offering many tools and commands to manage your Bot Framework environment. It can be installed using this link `https://aka.ms/bfcli` or by simply running the following command if you have Node.js already installed on your machine:

```
npm i -g @microsoft/botframework-cli
```

Currently, this CLI supports commands and management interfaces for the following services, which will be discussed in detail in the Chapters 4 and 5:

- **Chatdown**
  - *Chatdown* is a tool for parsing .chat files which are then translated into .transcript files. This makes the process of designing dialogs and conversations easier, without writing actual code, and is therefore used in the design phase of a bot project

- **QnAMaker**
  - Using the *QnAMaker* commands from the BF CLI, developers can manage everything related to QnA Maker, like creating new knowledge bases or training and publishing existing ones using the CLI instead of the QnA Maker portal.

- **Config**
  - The *Config* endpoints within the CLI let you manage the config key-value pairs for your environment, like setting the necessary LUIS keys and IDs which are required by other commands, which streamlines the execution of commands.

- **LUIS**
  - The *LUIS* commands offer management interfaces and features to handle your LUIS applications used within your bot. With these commands, you could simply create, train, and publish a new language understanding application from the CLI, which allows you to automate that process, while developing a bot.

## Adaptive Cards

Adaptive Cards is an open-source concept which allows card authors to describe rich attachment cards using either JSON or the visual designer available via `https://adaptivecards.io/designer/`. These cards are then rendered natively within the channel the bot is deployed to, making it easier for developers to build multichannel bots as there is no need of covering the user interface parts for each piece separately. The main idea is to use Adaptive Cards to bundle text, images, buttons, and other rich media types in a single message, which is displayed in the native look and feel of the used channel. Such an Adaptive Card could, for example, look like the one shown in Figure 2-12 when displayed in the Bot Framework Web Chat.

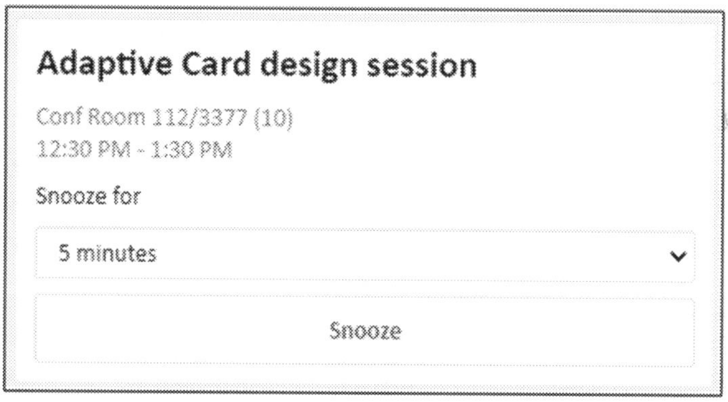

Adaptive Card design session
Conf Room 112/3377 (10)
12:30 PM - 1:30 PM
Snooze for
5 minutes
Snooze

***Figure 2-12.*** *Adaptive Card BF Web Chat example*

Figure 2-13 shows how the same card would look like if it was rendered in Microsoft Teams.

Teams light mode
Teams dark mode
Test Bot 2:36 PM
Adaptive Card design session
Conf Room 112/3377 (10)
12:30 PM - 1:30 PM
Snooze for
5 minutes
Snooze

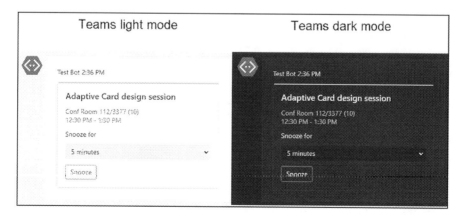

***Figure 2-13.*** *Adaptive Card Microsoft Teams example*

The card, as earlier mentioned, is only declared in JSON, which offers the possibility to embed it in multiple channels without changing the card definition. The JSON used for the preceding card is the following:

```json
{
    "$schema": "http://adaptivecards.io/schemas/adaptive-card.json",
    "type": "AdaptiveCard",
    "version": "1.0",
    "speak": "Your  meeting about \"Adaptive Card design session\" is
    starting at 12:30pmDo you want to snooze  or do you want to send a late
    notification to the attendees?",
    "body": [
      {
        "type": "TextBlock",
        "text": "Adaptive Card design session",
        "size": "large",
        "weight": "bolder"
      },
      {
        "type": "TextBlock",
        "text": "Conf Room 112/3377 (10)",
        "isSubtle": true
      },
      {
        "type": "TextBlock",
        "text": "12:30 PM - 1:30 PM",
        "isSubtle": true,
        "spacing": "none"
      },
      {
        "type": "TextBlock",
        "text": "Snooze for"
      },
```

```
  {
    "type": "Input.ChoiceSet",
    "id": "snooze",
    "style": "compact",
    "value": "5",
    "choices": [
      {
        "title": "5 minutes",
        "value": "5"
      },
      {
        "title": "15 minutes",
        "value": "15"
      }
    ]
  }
],
"actions": [
  {
    "type": "Action.Submit",
    "title": "Snooze",
    "data": {
      "x": "snooze"
    }
  }
]
}
```

Table 2-6 lists the currently supported channels and host applications for Adaptive Cards including Microsoft applications as well as third-party channels.

***Table 2-6.***  *Adaptive Cards channel support status (source:* `https://docs.` `microsoft.com/en-us/adaptive-cards/resources/partners`*)*

| Platform | Description | Version |
|---|---|---|
| **Bot Framework Web Chat** | Embeddable web chat control for the Microsoft Bot Framework. | 1.2.3 (Web Chat 4.7.1) |
| **Outlook Actionable Messages** | Attach an actionable message to email. | 1.0 |
| **Microsoft Teams** | Platform that combines workplace chat, meetings, and notes. | 1.2 |
| **Cortana Skills** | A virtual assistant for Windows 10. | 1.0 |
| **Windows Timeline** | A new way to resume past activities you started on this PC, other Windows PCs, and iOS/Android devices. | 1.0 |
| **Cisco WebEx Teams** | Webex Teams helps speed up projects, build better relationships, and solve business challenges. | 1.2 |

# Introducing Bot Framework Composer

The Bot Framework Composer is a visual integrated development environment for building bots without writing actual C# or JavaScript code. As of today, it is still in preview, which has been announced at Microsoft Ignite 2019. The following should outline the key points of this tool, as later chapters will go into detail on how to leverage Composer to build and deploy Bot Framework bots.

As Figure 2-14 shows, the Bot Framework Composer offers a visual designer which is built to create and manage dialogs in a simple graphical way. Moreover, the Composer has built-in functionality to create and maintain natural language models, which avoids the need to jump between different tools and portals while developing a bot. Besides the natural language aspect, it also provides a language generation engine to generate more sophisticated and natural conversations. Additionally, you can run your created bot right from Composer which makes testing and debugging even easier for developers as well as business users.

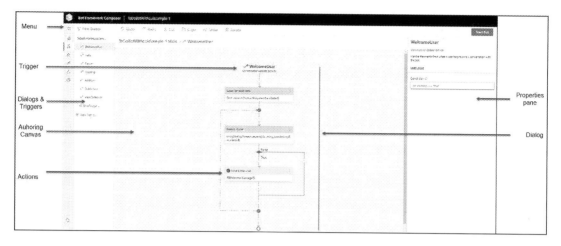

**Figure 2-14.**  *Bot Framework Composer overview*

# Advantages of Composer

One of the key points of Composer is that is intended to be used by developers and nondevelopers. This tool should bridge the gap between engineering and business when working on a conversational AI application together, as business users can use Composer to outline the conversation design and hand it off to developers then to implement functionality which requires code.

Furthermore, Composer is basically writing the bot logic in the form of JSON and markdown for the user, which can then be extended and adopted to fulfill the requirements. Thus, the reusable components created by Composer can be source-controlled along with other parts of the bot like language understanding definition to build a holistic bot project in a team of developers and nondevelopers.

Composer utilizes and integrates many different components of the conversational AI platform in a single interface including the following:

- Bot Framework SDK

- Adaptive dialogs

- Language Understanding service using LUIS

- Language Generation

- QnA Maker

- Bot Framework Emulator

63

# Adaptive Dialogs

In contrast to the current concept of the Bot Framework SDK, Composer relies on a concept called "Adaptive Dialogs" to build dialogs within a bot. This concept basically lets you define dialogs in JSON using a declarative approach instead of C# or JavaScript code. This way, adaptive dialogs can be exchanged during runtime, making the deployment and integration process of a Bot Framework bot more reliable and faster.

Adaptive dialogs can be somehow seen as a new way of modeling and implementing dialogs in conversational applications as it helps focusing on the conversation modeling rather than the implementation requirements for managing dialogs.

# Language Understanding

As language understanding is a key component in bot design and development, Composer offers a simplified editor for managing natural language understanding models in the context of a dialog using a markdown expression language. This enables users to build and extend dialogs along with language understanding models in a single interface. Within this markdown-like format, you can define intents, utterances, and entities, which you would normally do in the LUIS portal. These language understanding models are stored in .lu files, making it easier to coauthor language models. The file format is outlined in Figure 2-15.

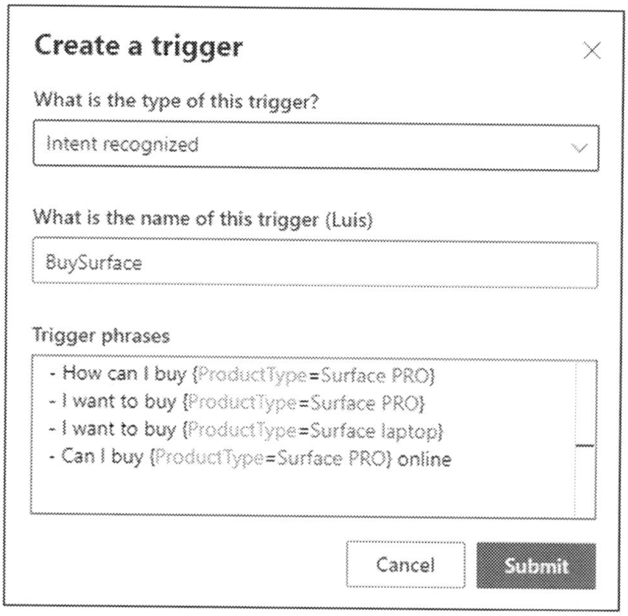

**Figure 2-15.** *Bot Framework Composer language understanding example*

# Language Generation

In addition to the language understanding integration, Composer also offers a simple way to integrate language generation into your bot. Similar to the language understanding markdown-like format, you can define language generation models in using markdown as well, as the following illustration shows. The bot will then use the language generation models, stored as .lg files, to randomly pick one phrase from it before sending it to the user. This makes the conversation more natural and sophisticated as the bot does not respond to the same input the same way each and every time. The format of such a .lg file is outlined in Figure 2-16.

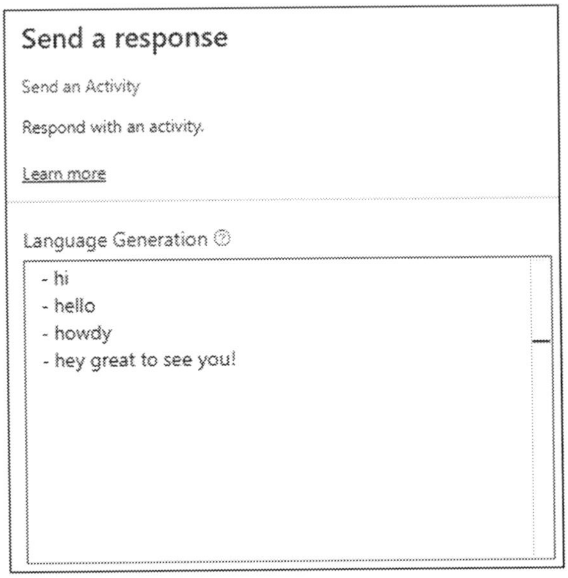

**Figure 2-16.**  *Bot Framework Composer language generation example*

# Summary

In this chapter you learned all key concepts like activity processing or dialog management within the Microsoft Bot Framework. Furthermore, we looked at a typical bot project structure and went through the code of a newly created bot using the Microsoft Bot Framework SDK for C# and JavaScript. Additionally, we discussed the tools and offerings around the Bot Framework which are needed to develop, test, and maintain bots built on top of the conversational AI platform.

In the next chapter, the Azure Cognitive Services are covered in more detail focusing on the main categories which are used in bot projects. Moreover, some of the best practices and guidelines from the field will be included, outlining how to combine these Cognitive Services with bots built using the Microsoft Bot Framework SDK.

# CHAPTER 3

# Introduction to Azure Cognitive Services

Conversational apps and agents usually stand out from other applications by their intelligent capabilities. These intelligent skills can usually be categorized into the following categories:

- Language

- Speech

- Vision

- Decision

Within each of these categories, there can be found a lot of different use cases which a chatbot or conversational app could benefit from, may it be the ability to understand messages sent by users in the form of text or the competence to turn spoken words into text for further processing. Microsoft provides a set of APIs called Azure Cognitive Services, allowing developers to add intelligence to applications. The overall goal is to let developers infuse any kind of cognitive abilities into applications to make those apps smarter and more human-alike. The architecture of these Cognitive Services allows users to combine two or more Cognitive Services with each other, which is especially useful in chatbot use cases.

Therefore, this third chapter of the book deals with many of the intelligent tools mentioned in Chapter 1, which are somewhat important for building a successful conversational AI application using the Microsoft Azure platform. The focus, however, is clearly on the language category, as those are the most important ones, when building a conversational application.

© Stephan Bisser 2021

S. Bisser, *Microsoft Conversational AI Platform for Developers*, https://doi.org/10.1007/978-1-4842-6837-7_3

# Language Category: Extract Meaning from Unstructured Text

The language category is one of the most important aspects in a conversational application. When building an application, like a chatbot, which is designed to interact with users in a conversational manner, it is crucial to infuse language-driven intelligent skills into the app. While language understanding could be one aspect to that, there are other facets which deal with language in a conversational app, like text analytics or text translation. Thus, the following section should describe the concepts behind some of those Cognitive Services in the language category in detail.

## Language Understanding (LUIS)

As stated in Chapter 1, understanding the natural text language within any conversational application is an essential part and will decide if the application will be used by people or not. Therefore, Microsoft offers an API called Language Understanding (shortened with LUIS in the following text) which offers natural language processing capabilities. The idea behind LUIS is to empower people with a custom natural language understanding model, without the needs of developing the necessary algorithms themselves:

> *Language Understanding (LUIS) is a cloud-based API service that applies custom machine-learning intelligence to a user's conversational, natural language text to predict overall meaning, and pull out relevant, detailed information.*
>
> *A client application for LUIS is any conversational application that communicates with a user in natural language to complete a task. Examples of client applications include social media apps, chat bots, and speech-enabled desktop applications. (Microsoft Docs, 2020)*

### LUIS Building Blocks

The key components of a LUIS app, which are vital to understand how this service works, are the following.

***Table 3-1.*** *LUIS components*

| Component | Description |
| --- | --- |
| Utterance | The utterance is the unstructured input text, which is sent by the user to the conversational application (e.g., a chatbot). An example for that would be the sentence "What is the weather like in Seattle?" |
| Intent | An intent is the definition of a concrete task, which the user wants to execute. It can be seen as the aim or objective expressed by the user. Given the utterance "What is the weather like in Seattle?" the intent of this utterance would be *GetWeather*. |
| Entity | Entities are text parts within utterances which can be used to derive certain data from the unstructured text, like important values such as a name or a date. The entity of the abovementioned utterance within the *GetWeather* intent would be, for example, the location, like in this case *Seattle*. |
| Score | The score represents how confident the LUIS app is that a given utterance matches a certain intent or an entity within an utterance has been identified. The score is a number between 0 and 1, whereas a prediction score close to 1 constitutes a relative high confidence and a score close to 0 means that LUIS has low confidence that the result is accurate or correct. |

That means from a practical point of view that whenever you invoke a LUIS application via its API endpoint to gather the intent and the entity/entities of a given user utterance, you'll get the result in the form of intents and entities which have been detected, as shown in Figure 3-1, back.

***Figure 3-1.*** *LUIS basic explanation*

From a development perspective, the result you will get back is a JSON representative. So, you basically get the following JSON string back as the result to the utterance mentioned in Figure 3-1:

```
{
    "query": "What is the weather like in Seattle.",
    "topScoringIntent": {
        "intent": "GetWeather",
        "score": 0.921233
    },
    "entities": [
        {
            "entity": "Seattle",
            "type": "Location",
            "score": 0.9615982
        }
    ]
}
```

Consequently, LUIS is not exclusively designed to be used in chatbot use cases only but can also be part of other applications, where language understanding is a necessity. However, as this book covers the conversational AI platform, the following text will deal with LUIS in conversational use cases.

## Creating a LUIS Application

Before you can create a LUIS application, you would need to make sure that you have access to a valid Microsoft Azure subscription. It is important that your account can create new resources in that Azure subscription; otherwise, you will not be able to complete the following steps.

---

**Note**   If you do not already have an active Azure subscription in use, you can sign up for a new Azure subscription with 200$ for free using this link `https://azure.microsoft.com/en-us/free`.

---

There are mainly two different ways of creating a new LUIS application, using the web-based portal and using the CLI. Both approaches will be demonstrated in the following paragraphs. The outcome of both options is the same, although, using the CLI, you will get a higher chance of automating these tasks. This is extremely helpful in cases where you need to create many LUIS applications at a time or update many LUIS applications simultaneously.

Before you can provision a new LUIS application, you would need to create a new Azure resource group in your Azure subscription. To complete that, log in to `https://portal.azure.com` and create a new resource group as shown in Figure 3-2.

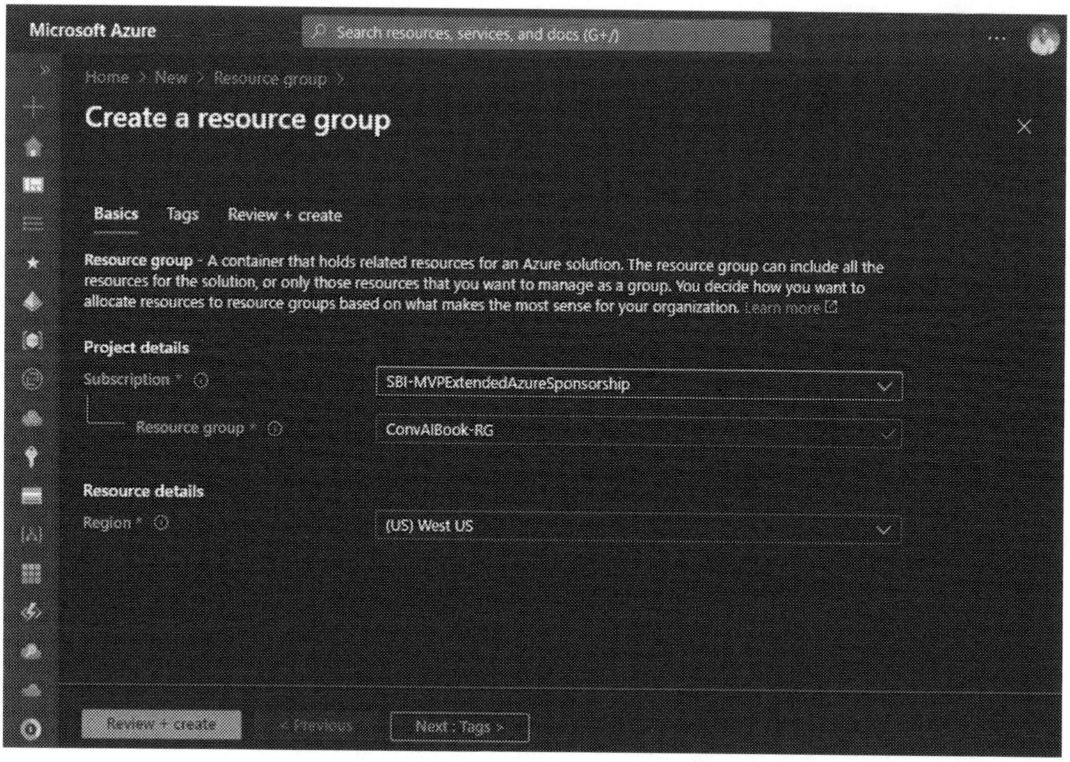

*Figure 3-2.*  *Create a new Azure resource group for LUIS*

## Using the LUIS Portal

In order to create a new LUIS application, you need to go to `www.luis.ai/` and sign in with your account which has access to the Azure subscription. After signing in, click your name from the top right corner, and select "Settings" as shown in Figure 3-3.

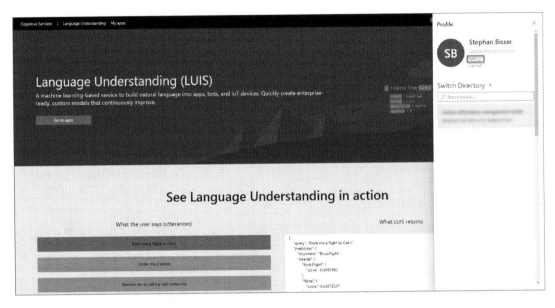

***Figure 3-3.*** *Set up LUIS application 1 – log in*

From the settings screen, you have the option to manage your account and the respective settings. In our case, we need to create an authoring resource, meaning an Azure resource responsible for authoring LUIS applications. So, the next step is to create a new authoring resource by clicking the corresponding button as illustrated in Figure 3-4.

***Figure 3-4.*** *Set up LUIS application 2 – user settings*

Now you need to enter the resource name and select the Azure subscription as well as the previously created Azure resource group in order to create a new LUIS authoring resource, as outlined in Figure 3-5.

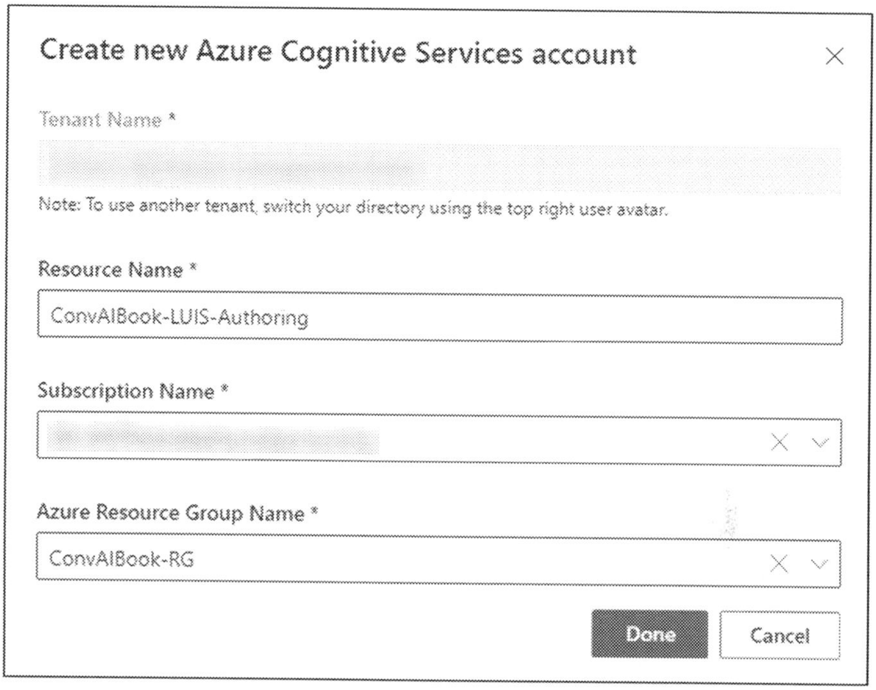

**Figure 3-5.** *Set up LUIS application 3 – create new authoring resource*

Going back to `www.luis.ai/applications` and selecting your Azure subscription and your authoring resource, you should see a similar screen as in Figure 3-6, as you do not have any LUIS applications created yet.

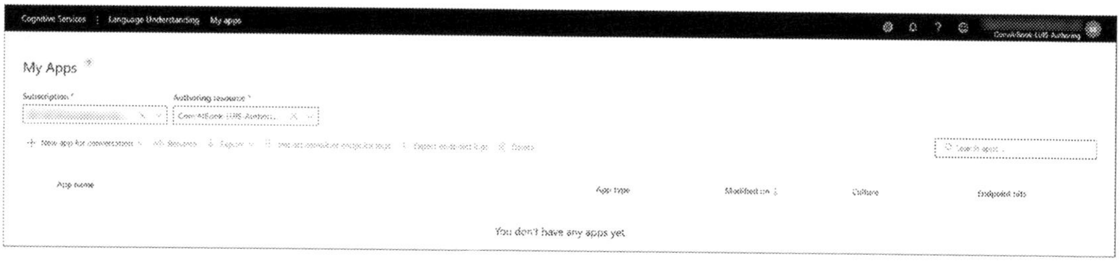

**Figure 3-6.** *Set up LUIS application 4 – My apps view*

Now you can go ahead and create a new LUIS application for conversation. Basically, you have three options of creating a new LUIS application:

- **New app for conversation**: This option means you create a new LUIS model from scratch.

- **Import as JSON**: This option means you import an already JSON existing LUIS model.

- **Import as LU**: This option is like "Import as JSON," but instead of importing a JSON representative of the LUIS model, you import a .lu file, which is typically easier to read compared to JSON files.

---

**Note**   .lu files are markdown-like files which describe or define a LUIS application. In general a .lu file consists of the definition of intents along with the intents' entities and utterances.

---

For the sake of simplicity, you should go with the first option "New app for conversation" to start from scratch with the language understanding application, like shown in Figure 3-7.

***Figure 3-7.***  *Set up LUIS application 5 – create new LUIS app*

The last step before the LUIS application gets created is to insert the app's information, like the name and the culture to be used. Assign your LUIS application a descriptive name, and pick the culture you want your language understanding model to be based on, as outlined in Figure 3-8.

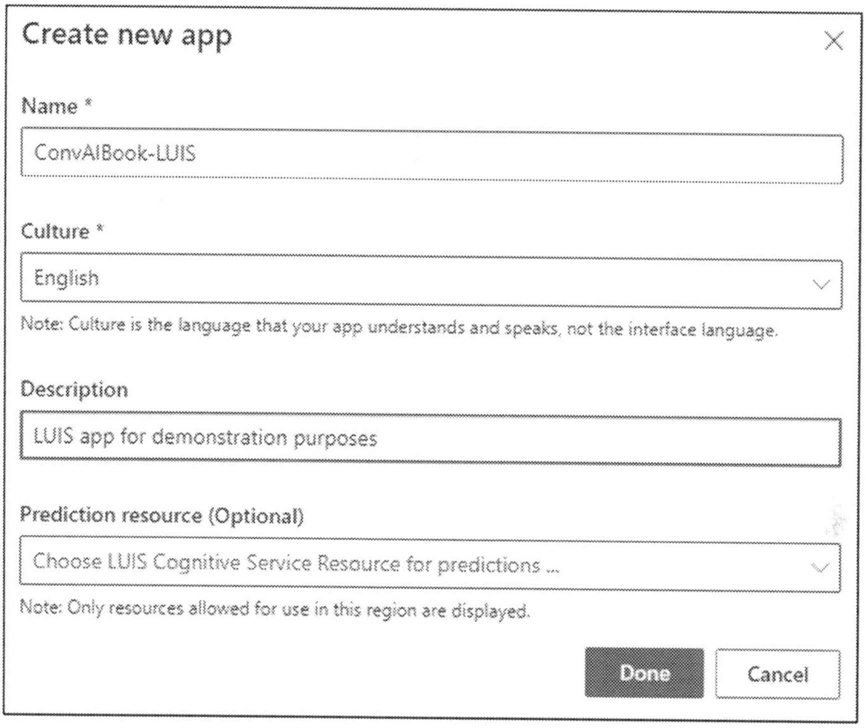

**Figure 3-8.** *Set up LUIS application 6 – insert LUIS app's details*

Now that the LUIS application has been created, the next step is to create the need intents and entities, which should be used in your LUIS app. The goal is to create a language understanding model, as outlined in Figure 3-9, which consists of two intents. The first intent is used to handle requests covering the weather, whereas the second intent covers booking a table at a restaurant.

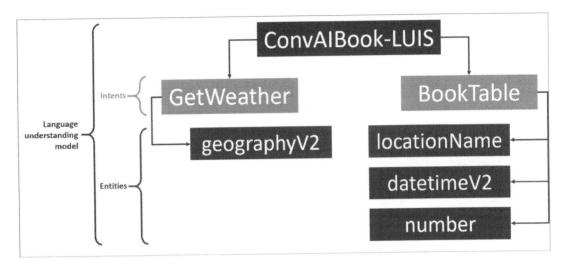

**Figure 3-9.** *Sample LUIS model structure*

As the LUIS application is now created, you can move on and alter the app to fit your use case. When viewing your newly created LUIS app, you will notice that there is already one intent created for you, the *None* intent. This intent's purpose is to act as a fallback intent within your LUIS application and should consist of utterances which are outside of your language model or schema. First, you will need to add the necessary intents to your application, *GetWeather* and *BookTable*, as shown in Figure 3-9. This can be done by clicking the "+ Create" button on the Intents page and type in the intents' names, as outlined in Figure 3-10.

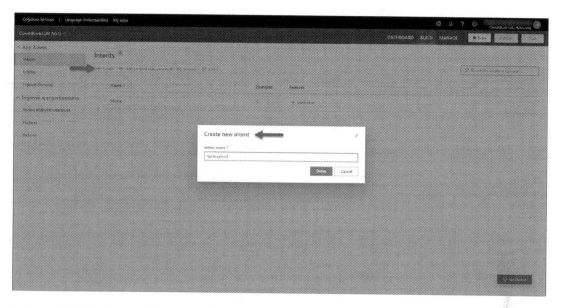

*Figure 3-10.  Set up LUIS application 7 – create intents*

---

**Note**    LUIS also offers so-called prebuilt domain intents which allow you to add popular intents to your application which come with a set of predefined utterances and entities covering common scenarios.

---

The next step is to add the necessary entities to our LUIS application, which are geographyV2 (a prebuilt entity targeting to predict place names, like cities or countries), *locationName* (a machine-learned entity designed to predict location names, like restaurants), *datetimeV2* (a prebuilt entity targeting to predict date time values), and *number* (a prebuilt entity targeting to predict numbers, which is used to predict the number of people for bookings), according to Figure 3-9. Here we are going to take advantage of the some of the prebuilt entities LUIS offers. So, you need to add the prebuilt entities mentioned in that figure to your LUIS app. As there is one entity left, which is not part of the prebuilt entities, you need to create another entity called *locationName* of type "Machine learned," as demonstrated in Figure 3-11.

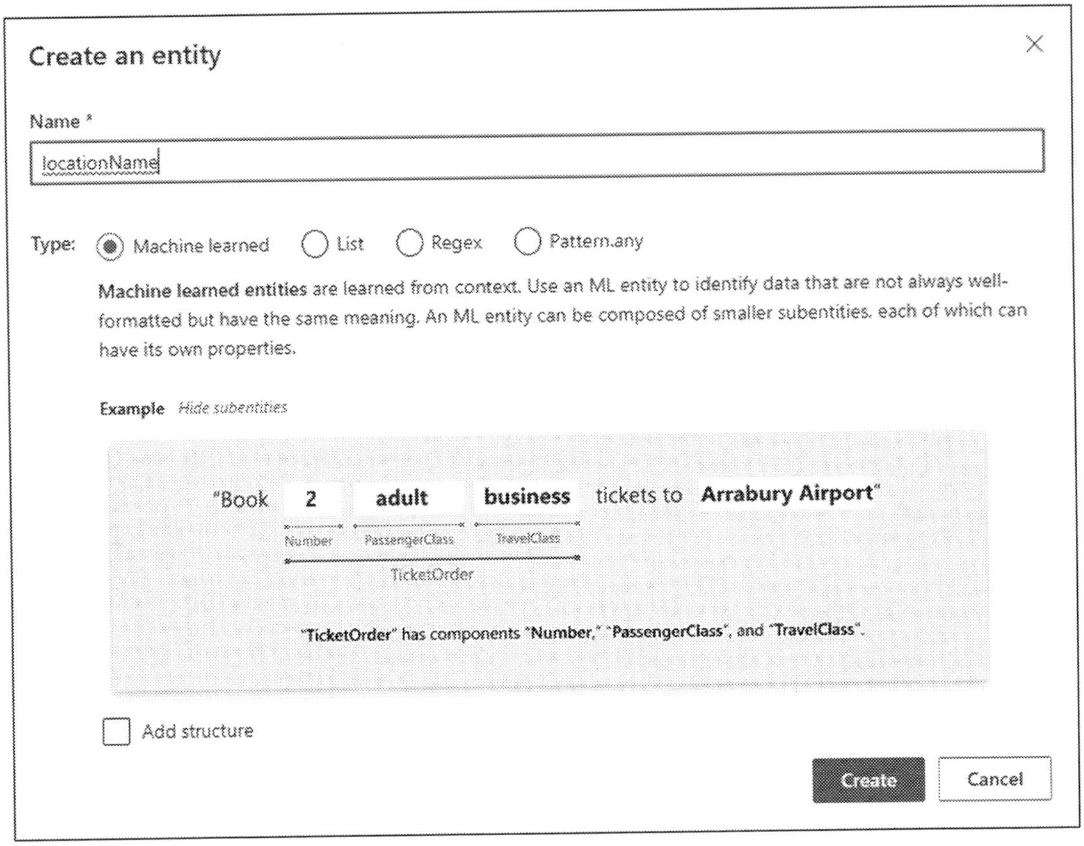

**Figure 3-11.** *Set up LUIS application 8 – create entities*

After adding the intents and entities, you need to add sample data, called *utterances*, to your LUIS model. The concept behind utterances is to tell your LUIS application what kind of user input it should detect and how to predict certain values within those user input phrases correct. Therefore, it is also a good approach to add a mixture of distinct utterances to your LUIS application, to gain better prediction results. So, for the GetWeather intent, you could add the following utterances to your app, for example:

- What is the weather like in Seattle?

- How is the weather in New York?

- What's the weather in Berlin'?

- Could you tell me the weather forecast for Redmond?

- I would like to get some details on the weather in Amsterdam please.

After adding these utterances to your GetWeather intent, you should immediately see that LUIS will mark city names with the geographyV2 entity, as shown in Figure 3-12. This means that you do not need to tag the city names within the utterances by yourself, to tell LUIS that those are city names, which should be extracted.

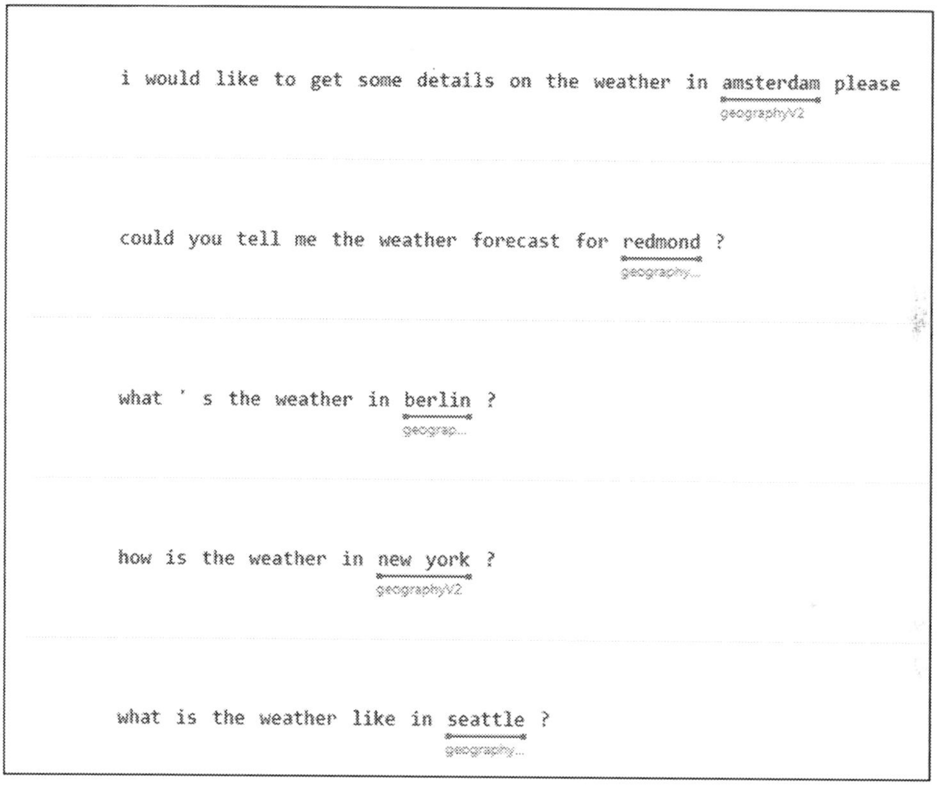

*Figure 3-12.* *Set up LUIS application 9 – add utterances for GetWeather*

Now you need to add some utterances to the BookTable intent as well. You could, for example, add the following utterances to your intent:

- Can you reserve a table for four people at the famous sushi bar tomorrow?

- Reserve a table for Saturday 3 p.m. at Hard Rock Café for six people please.

- Please reserve a spot for five at Jamie's kitchen on 23rd of November.

- Could you reserve a table at Tom's diner tomorrow?

- Book a table at the Redmond Steak House and Grill please.

What you should see is that LUIS is able to detect the number of people as well as the dates and times correctly, but the location names will not be extracted at all, as seen in Figure 3-13. This is due to the fact that you have not advised your LUIS app what a location name looks like.

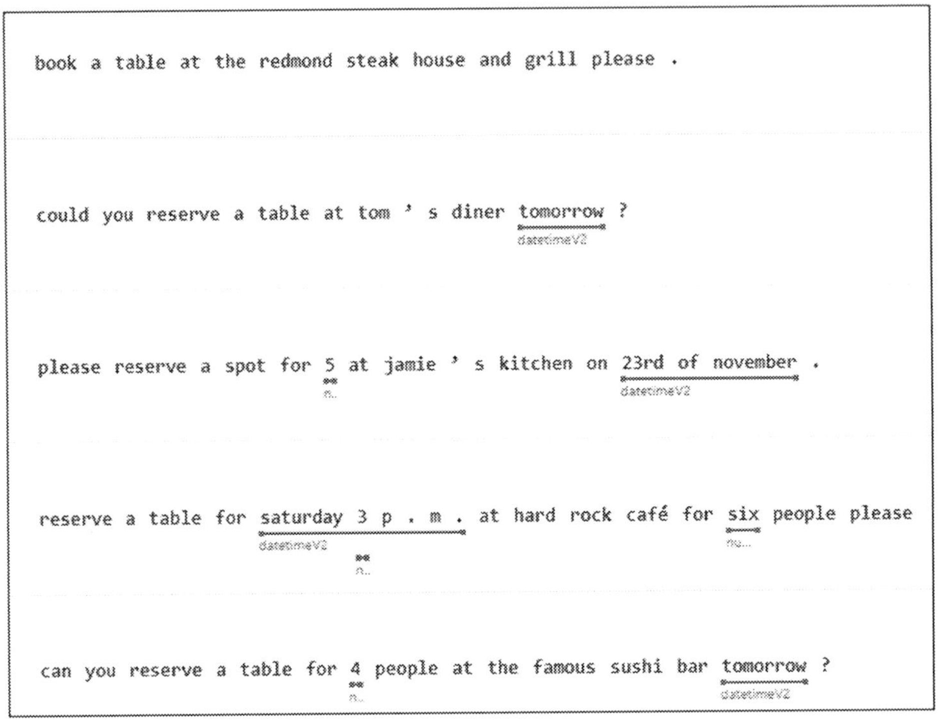

*Figure 3-13.* *Set up LUIS application 10 – add utterances for BookTable*

However, you can easily mark the location names within the utterances as entities, to tell LUIS what an entity looks like. To do that, simply hover over a word within an utterance and click it. If the entity consists of more than one word, simply hover over and click the first word and then click the last word within the entity representation, as shown in Figure 3-14.

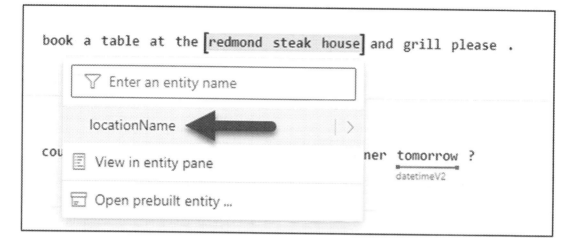

book  a  table  at  the [redmond  steak  house] and  grill  please  .

▽  Enter an entity name

locationName                                    ⟩

cou  ▤  View in entity pane                ner  tomorrow  ?
                                                    datetimeV2

□  Open prebuilt entity ...

**Figure 3-14.** *Set up LUIS application 11 – mark entity*

Mark all the locationName entities within the utterances, to tell LUIS that these kinds of words or phrases represent a location name, as outlined in Figure 3-15.

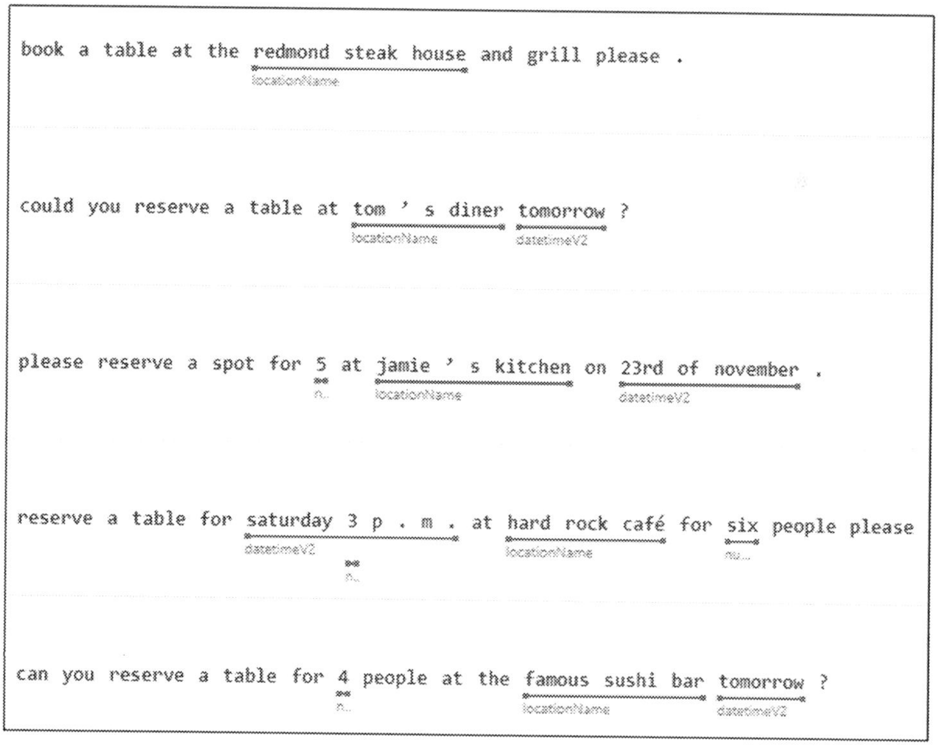

book  a  table  at  the  redmond  steak  house  and  grill  please  .
                         locationName

could  you  reserve  a  table  at  tom  ' s  diner  tomorrow  ?
                                   locationName       datetimeV2

please  reserve  a  spot  for  5  at  jamie  ' s  kitchen  on  23rd  of  november  .
                               n..    locationName           datetimeV2

reserve  a  table  for  saturday  3  p  .  m  .  at  hard  rock  café  for  six  people  please
                        datetimeV2                  locationName        nu...
                                  n..

can  you  reserve  a  table  for  4  people  at  the  famous  sushi  bar  tomorrow  ?
                                  n..                 locationName        datetimeV2

**Figure 3-15.** *Set up LUIS application 12 – BookTable marked entities*

Now that you have added all intents, entities, and utterances, you need to train your LUIS application. This process tells your LUIS app to improve or enhance the level of language understanding. Typically, you would train your LUIS application after adding new intents, entities, or utterances as well as after editing certain parts of your LUIS app to update it. The process of training your app is rather simple, as shown in Figure 3-16; the only thing you need to do is to click the "Train" button from the upper right corner of the LUIS portal, which starts the training process.

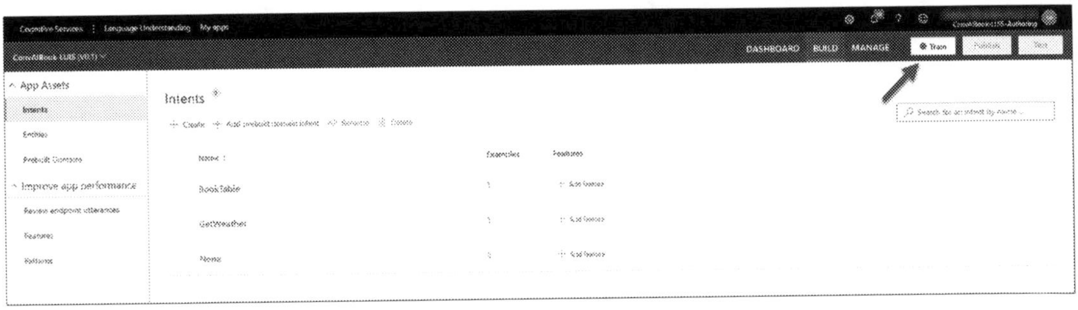

***Figure 3-16.*** *Set up LUIS application 13 – train LUIS app*

After the training of your LUIS app has been completed, you will receive a notification in your notifications list, stating that the training has been finished successfully and all changes are now effective, like demonstrated in Figure 3-17.

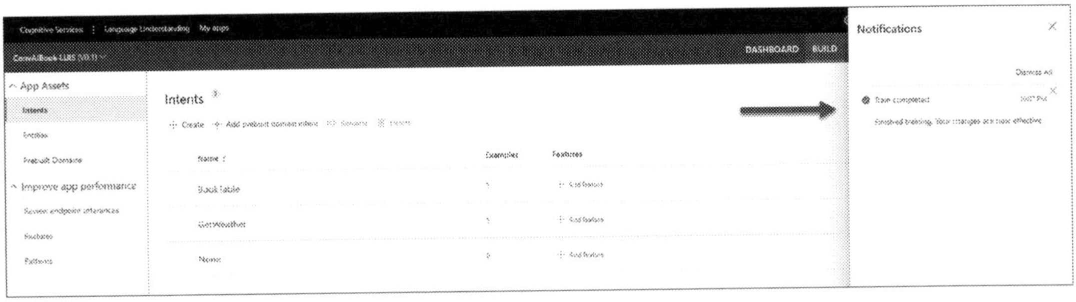

***Figure 3-17.*** *Set up LUIS application 14 – notifications after training*

After the training process of your application has been finished, it is time to test the language model for the first time. To do so simply open up the testing pane as illustrated in Figure 3-18 from the top right corner of the LUIS portal, and insert sample phrases which are not part of the list of utterances.

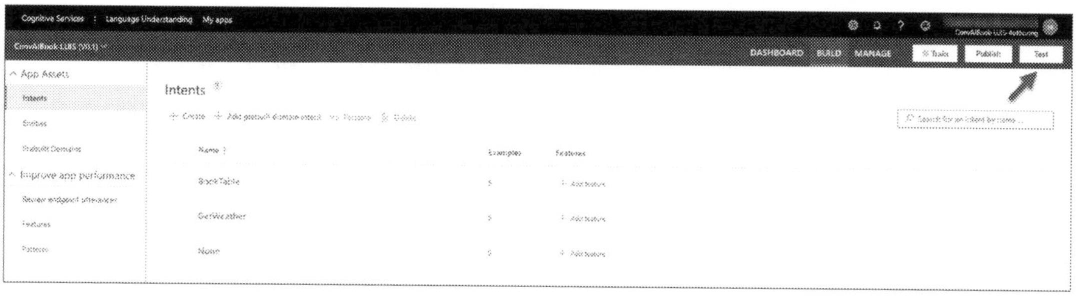

***Figure 3-18.*** *Set up LUIS application 15 – test*

For instance, you could use the following phrase to test the BookTable intent, which should give you the result as shown in Figure 3-19. The app has detected that the phrase *"Could you book a table at joey's restaurant for 7 on Sunday please?"* belongs to the intent BookTable confidence score of 0.872. Additionally, the application has extracted the entities "joey's restaurant" as location name, "7" as the number of people, and "Sunday" as date-time correctly. This demonstrates that you do not need to add a lot of sample data to your LUIS application before the language model will be sophisticated enough to fit your needs, but a little amount of utterances is already enough. Nevertheless, the more utterances you add to your app, the more precise your language model will be in the end.

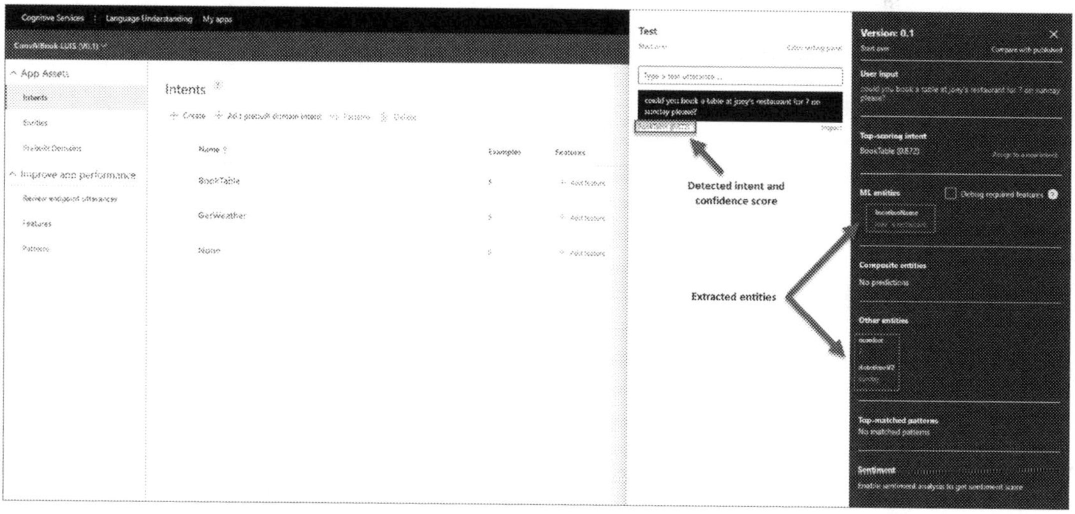

***Figure 3-19.*** *Set up LUIS application 16 – test results*

The last step in the process of setting up a LUIS application is the action of publishing the app. Upon publishing a LUIS app, as shown in Figure 3-20, the app becomes available within your LUIS API endpoint. This then lets you connect your client application, such as a chatbot, to your language model via the LUIS API, to retrieve intents based on inputs, for example. Publishing an app is typically done, after a major upgrade of the language model, when you want to expose your change to your client applications, or the first time when you finished setting up your language model.

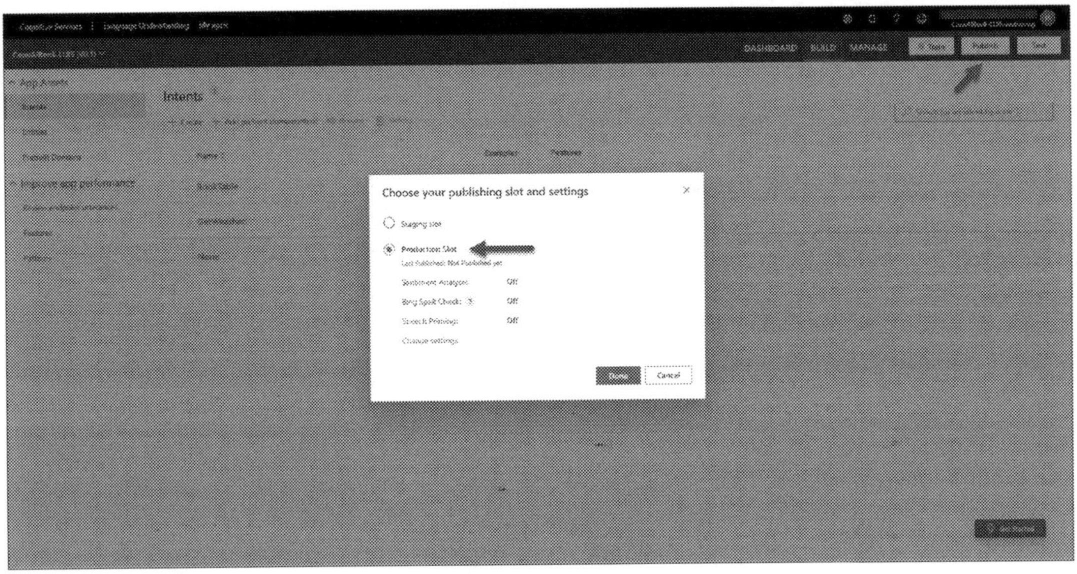

***Figure 3-20.*** *Set up LUIS application 17 – publish*

## Using the CLI

Although using the LUIS portal is a great way of learning the concepts and features of LUIS, there is also another way of managing a LUIS application, the Bot Framework CLI. As mentioned in Chapter 2, the Bot Framework CLI, which can be downloaded from https://aka.ms/bfcli, is designed to let Bot Framework developers manage all services related to a chatbot or conversational application using the command line. Therefore, the next part of this chapter demonstrates how you can leverage the Bot Framework CLI to create, update, test, and publish a LUIS application.

The basis for using the Bot Framework CLI for creating a new LUIS application is either a .lu file or a JSON file which describes the language understanding model. I think using a .lu file has the advantage that it is first of all better readable as well as easier to create compared to a JSON file. So, you need to create a new file in a folder of your choice with an appropriate name (this tutorial will use *03_convAIBook-LUIS** as a name for all LUIS files used) and the file name extension .lu, which represents a language understanding file. This .lu file should basically represent the LUIS application you created in the previous steps using the LUIS portal; therefore, it will contain the following parts:

- **LUIS app information**

- **Intent definitions**

- **Entity definitions**

The following represents a sample .lu file which describes the two intents *"BookTable"* and *"GetWeather"* along with their utterances and entities:

```
> LUIS application information
> !# @app.name = ConvAIBook-LUIS
> !# @app.desc = LUIS app for demonstration purposes
> !# @app.versionId = 0.1
> !# @app.culture = en-us
> !# @app.luis_schema_version = 7.0.0
> !# @app.tokenizerVersion = 1.0.0
> # Intent definitions

## BookTable
- book a table at the {@locationName=redmond steak house} and grill please.
- can you reserve a table for 4 people at the {@locationName=famous sushi
  bar} tomorrow?
- could you reserve a table at {@locationName=tom's diner} tomorrow?
- please reserve a spot for 5 at {@locationName=jamie's kitchen} on 23rd of
  november.
- reserve a table for saturday 3 p.m. at {@locationName=hard rock café} for
  six people please

## GetWeather
- could you tell me the weather forecast for redmond?
- how is the weather in new york?
- i would like to get some details on the weather in amsterdam please
- what is the weather like in seattle?
- what's the weather in berlin?
```

```
## None

> # Entity definitions
@ ml locationName

> # PREBUILT Entity definitions
@ prebuilt datetimeV2
@ prebuilt geographyV2
@ prebuilt number
```

If you copy all of the above-outlined parts into the newly created 03_convAIBook-LUIS.lu file and save it, you will be ready for the next step, which is converting the .lu file to a JSON file. The import command requires a JSON file to be passed as input parameter, which holds a definition of the LUIS application which should be created, so we need to execute the following command in a command line, where the Bot Framework CLI is installed:

```
bf luis:convert --culture "en-us" --in ".\03_convAIBook-LUIS.lu" --out
".\03_convAIBook-LUIS.json"
```

---

**Note**    Make sure that you choose the right culture for your LUIS application within the Bot Framework CLI commands if you intend to create a language model in a different language.

---

This command will generate a file called "03_convAIBook-LUIS.json" in the current folder for you, which defines the LUIS application you described in the .lu file before. This JSON file can now be imported into your LUIS account as a new LUIS app using the following command:

```
bf luis:application:import --endpoint "https://<region>.api.cognitive.
microsoft.com" --subscriptionKey "yourLUISSubscriptionKey" --name
"ConvAIBook-LUIS-CLI" --in ".\03_convAIBook-LUIS.json"
```

Please make sure to insert your LUIS subscription key, which can be obtained from the Azure portal in your LUIS azure resource and the correct LUIS endpoint region, which can also be found on the page where your subscription key can be seen, in order to run the command successfully. When the command has been executed successfully, you will see the ID for your newly generated LUIS application as in Figure 3-21. Make sure to copy the app ID and save it somewhere, as you will need it for later commands again.

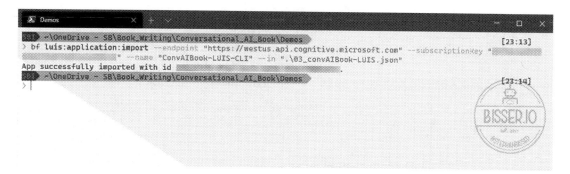

***Figure 3-21.*** *Import LUIS application using the Bot Framework CLI*

Now you can use the following command to view the details of your newly created LUIS application, like the description, the culture, the version, or the last time the app has been modified, which should lead the a similar output as in Figure 3-22:

```
bf luis:application:show --appId "appId" --endpoint "https://<region>.api.
cognitive.microsoft.com" --subscriptionKey "yourLUISSubscriptionKey"
```

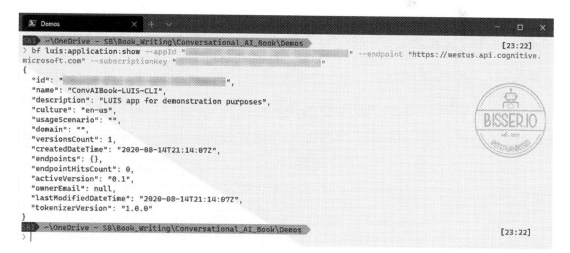

***Figure 3-22.*** *Show LUIS application information using the Bot Framework CLI*

The next command *bf luis:train:run* is used to train the LUIS application, before it can be tested. This command will tell the service that a new training request should be handled, so you need to execute the second command *bf luis:train:show* to track the training process of the application, as outlined in Figure 3-23:

```
bf luis:train:run --appId "appId" --versionId "0.1" --endpoint
"https://<region>.api.cognitive.microsoft.com" --subscriptionKey
"yourLUISSubscriptionKey"
```

```
bf luis:train:show --appId "appId" --versionId "0.1" --endpoint
"https://<region>.api.cognitive.microsoft.com" --subscriptionKey
"yourLUISSubscriptionKey"
```

*Figure 3-23.* *Train LUIS application using the Bot Framework CLI*

If all status indicators show "Success," the next step would be to test the LUIS application. This can easily be done by creating a new .lu file and adding test utterances for each intent which should be tested. This file which is called *03_luis_cli_test.lu* could, for example, look as follows:

```
# BookTable
- Could you book a table at joey's restaurant for 7 on Sunday please?
- I want to book a table for two at Jamie's kitchen tonight.
# GetWeather
- How's the weather in Vienna tomorrow?
- What's the weather like?
```

However, before you can test the LUIS app using the CLI, you need to publish it using the following command; otherwise, the test command would fail:

```
bf luis:application:publish --appId "appId" --versionId "0.1" --endpoint
"https://<region>.api.cognitive.microsoft.com" --subscriptionKey
"yourLUISSubscriptionKey"
```

Now to test the LUIS app, simply execute the following command and pass in the newly created *03_luis_cli_test.lu* as input file using the *-i* parameter, which should give you a similar result as shown in Figure 3-24:

```
bf luis:test -a "appId" -s "subscriptionKey" -i ".\03_luis_cli_test.lu"
```

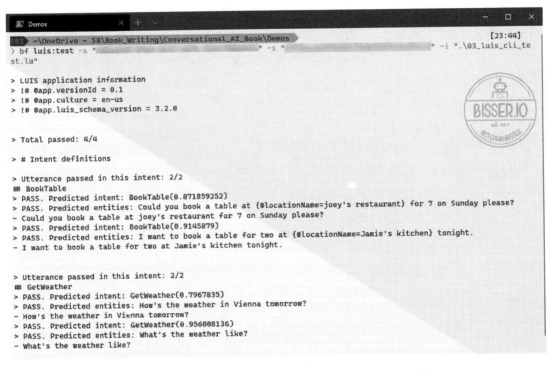

*Figure 3-24.* *Test LUIS application using the Bot Framework CLI*

**Note**   In production scenarios, you would need to add more than five utterances per intent to build up a sophisticated language model. Additionally, testing is the key to success, so make sure you include enough time within your project for testing and refining the language understanding model.

# QnA Maker

Another offering within the language category is a service called "QnA Maker." This is a tool designed to help you build sophisticated language models in a simple FAQ style. QnA Maker can be seen as knowledge base as a service, as it lets you build knowledge bases with your data with minimal effort. These knowledge bases can then be used by conversational applications, such as chatbots, to provide answers to users' questions. Hence, this section will cover all aspects which are essential for adding knowledge base data into a conversational application:

*Build, train and publish a sophisticated bot using FAQ pages, support web-
sites, product manuals, SharePoint documents or editorial content through
an easy-to-use UI or via REST APIs. (Microsoft, 2020)*

## QnA Maker Building Blocks

As Figure 3-25 illustrates, QnA Maker is dependent on different Azure services. Most
importantly is that all services and resources are deployed in your Azure subscription.
This means that the data stored and processed within your QnA Maker services is under
your control. QnA Maker itself uses a dedicated QnA Maker service, responsible for
holding your subscription keys. These are necessary for communicating with the QnA
Maker APIs.

*Figure 3-25.* *QnA Maker architecture*

All the data which you add into your QnA Maker knowledge base are stored in a dedicated Azure Search instance. Within that resource, the QnA pairs along with the associated metadata are stored and indexed. Additionally, you can store synonyms for certain words or phrases within your knowledge base, which is then also kept within Azure Search.

From a user perspective, you have the option to use the QnA Maker portal in order to create and maintain your knowledge bases. All operations which are taken within that portal are handled through the Azure App service, which is also part of your QnA Maker deployment. This resource holds the QnA runtime, responsible for handling API requests, like creating a knowledge base or adding a new QnA pair into your Azure Search store. The second part of the Azure App service is the QnA ranking component. This is heavily used when querying the QnA maker instance as it ranks the results which are retrieved by the Azure Search instance and determines the confidence scores before handing these results back to the client application.

The last service used in a typical QnA Maker deployment is Azure Application Insights. This cloud service stores all important telemetry data of your QnA Maker service which can be queried and analyzed. A common usage scenario for this is to analyze the top answered or unanswered questions based on user inputs.

Within a QnA Maker knowledge base, there are certain components which are essential to use as Figure 3-26 illustrates. Within a knowledge base, the QnA pairs are stored. Whenever your conversational application such as a chatbot queries the QnA Maker knowledge base, the input will be matched against the questions stored in the KB. If a match occurs, the answer is delivered as a response along with the confidence score of the match. Each QnA pair can also contain a metadata list. These key-value metadata pairs can be used to filter the results for a given query to get more accurate responses back from the QnA Maker service.

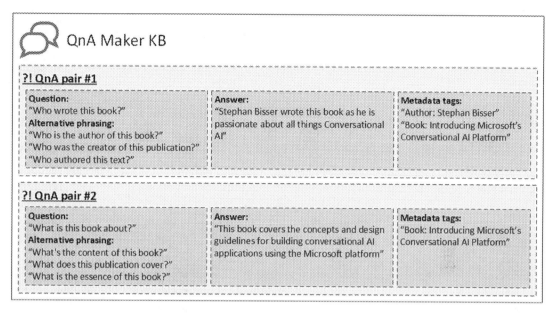

*Figure 3-26.* *QnA Maker KB structure*

# Creating a QnA Maker Service and Knowledge Base

There are two ways of creating a new QnA Maker instance along with a knowledge base, similar to LUIS, either using the GUI or the CLI. In this part, you will learn both ways, so you can either use the more graphical and end user–friendly way or use the CLI if you want to do it as part of a routine.

## Using the QnA Maker Portal

The QnA Maker portal can be accessed via `https://qnamaker.ai` using an account which has the permissions to create Azure resources. After you have logged in with your account, switch to "Create a knowledge base" from the top menu to start the creation process. From here you will first need to create a new QnA service under step 1. By clicking the link shown in Figure 3-27, you will be transferred to the Azure portal, where you might need to log in again with your account.

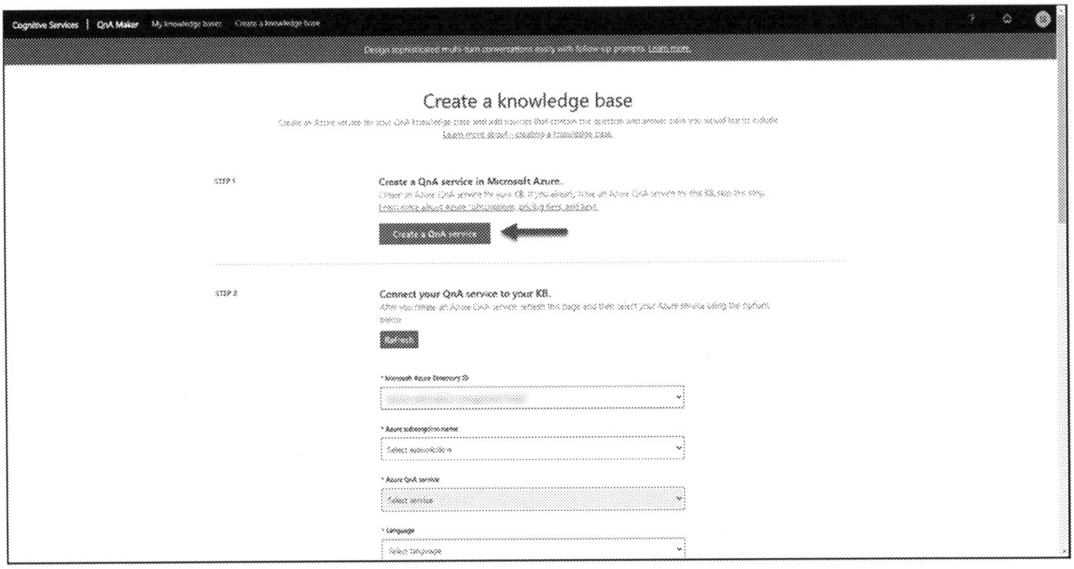

**Figure 3-27.**  *Set up QnA Maker KB 01*

After logging in, you will need to enter the information needed for creating your new QnA Maker Azure service, along with all other services which QnA Maker is dependent on (Azure Search, Azure App Service, and Application Insights), feel free to choose either the free pricing tiers or paid ones, but be aware that you can only provision one free QnA Maker instance per Azure subscription. So if you already have a QnA Maker service deployed in the Azure subscription you are using, you need to choose the Standard S0 tier, as shown in Figure 3-28.

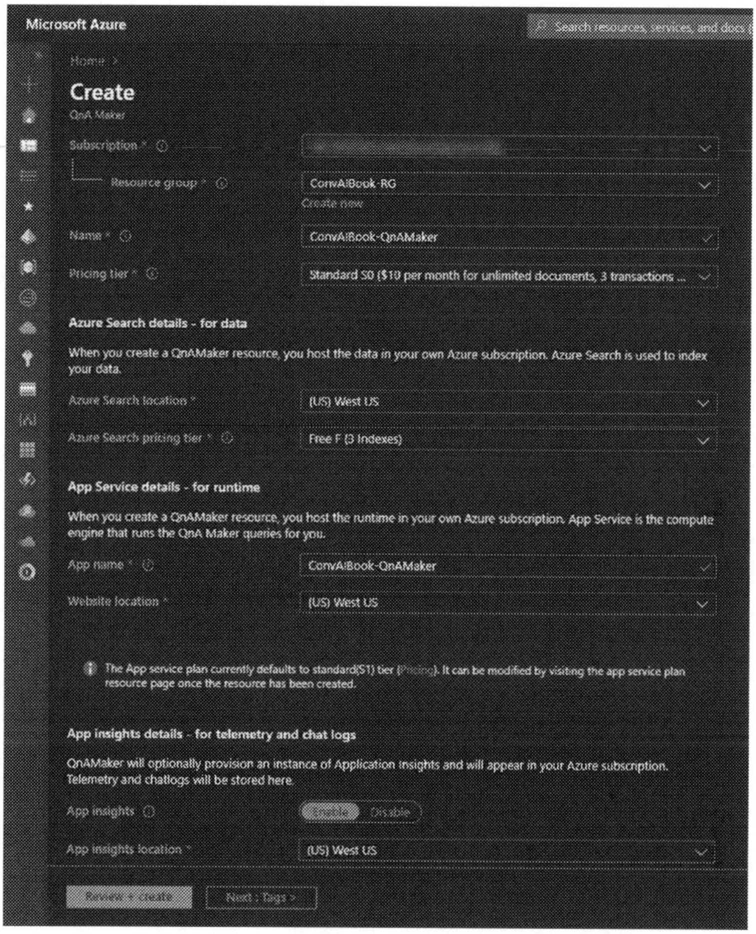

***Figure 3-28.*** *Set up QnA Maker KB 02 – create Azure resources*

After your Azure resources have been created, you can switch back to the QnA Maker portal and proceed with step 2. Within that step you need to select the right Azure Active Directory, the Azure subscription which holds your QnA service instance and the QnA service itself. Additionally, you need to choose the language of the knowledge base, as outlined in Figure 3-29. This is a crucial step, as you can only use one language per QnA service instance. So, by picking a language the first time you create a KB within your QnA service, the ranking algorithm will be deployed to your Azure service. This means you will need to create separate Azure QnA service instances for different languages if you want to build multilingual applications.

95

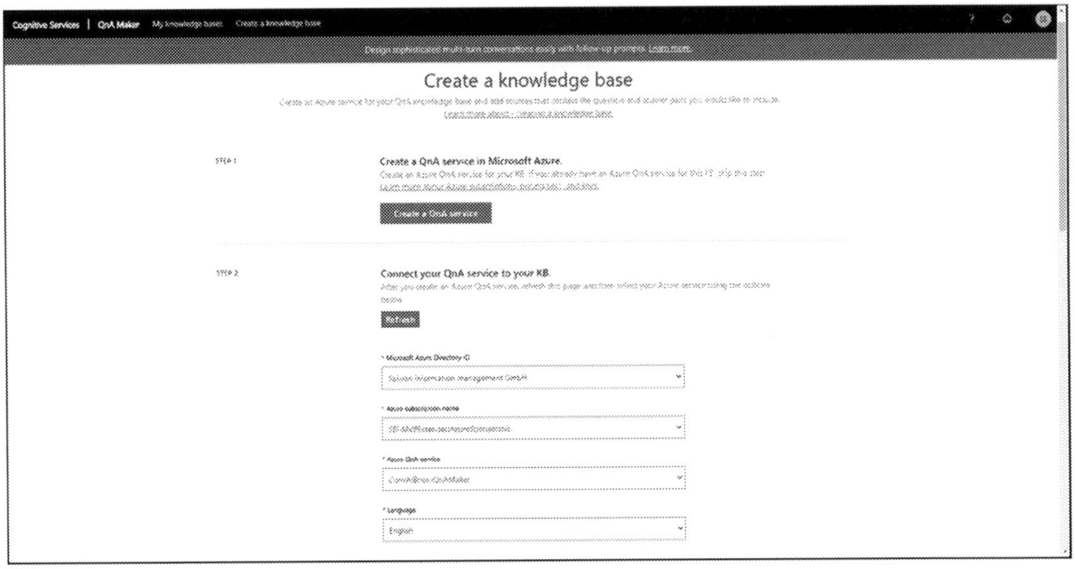

**Figure 3-29.** *Set up QnA Maker KB 03 – connect your QnA service to your KB*

---

**Note**   One QnA Maker Azure service instance can only hold knowledge bases of the same language. So if you plan to use multiple languages within your KBs, you need to deploy one QnA Maker service along with all services used for QnA Maker like Azure Search and Azure App Service for each language you plan to use.

---

In step 3 you need to give your knowledge base a name, which you can change at any time. In step 4 you can already populate your KB with content either from existing websites by entering URLs. Alternatively, you can upload FAQ-based files of the following types:

- PDF

- DOC

- Excel

- TXT

- TSV

In addition, you can also turn on the multiturn extraction feature, which also extracts the hierarchy of question and answer pairs of a given document or website. This then adds what is called "follow-up prompts" to a QnA pair, meaning that if, for example, a user asks a certain question which has a follow-up prompt attached to it, the application would represent the answer as well as the "child" question which is attached to it as a prompt, allowing users to quickly move on within the conversation. And the last step before you can create your knowledge base is to choose from one of the five prebuilt personality types, which populates your KB with chit-chat question and answer pairs as illustrated in Figure 3-30.

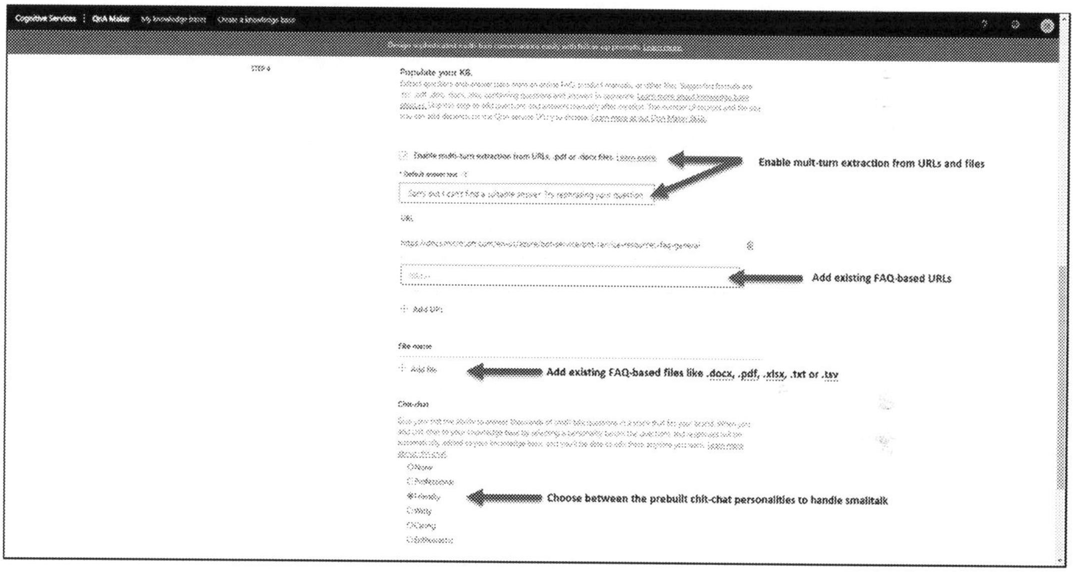

***Figure 3-30.*** *Set up QnA Maker KB 04 – add URLs, files, and chitchat*

After the creation of your knowledge base has been successful, your knowledge base is ready to be populated with your questions and answers. If you added one of the prebuilt chit-chat personalities, your knowledge base has already a lot of QnA pairs included, as shown in Figure 3-31.

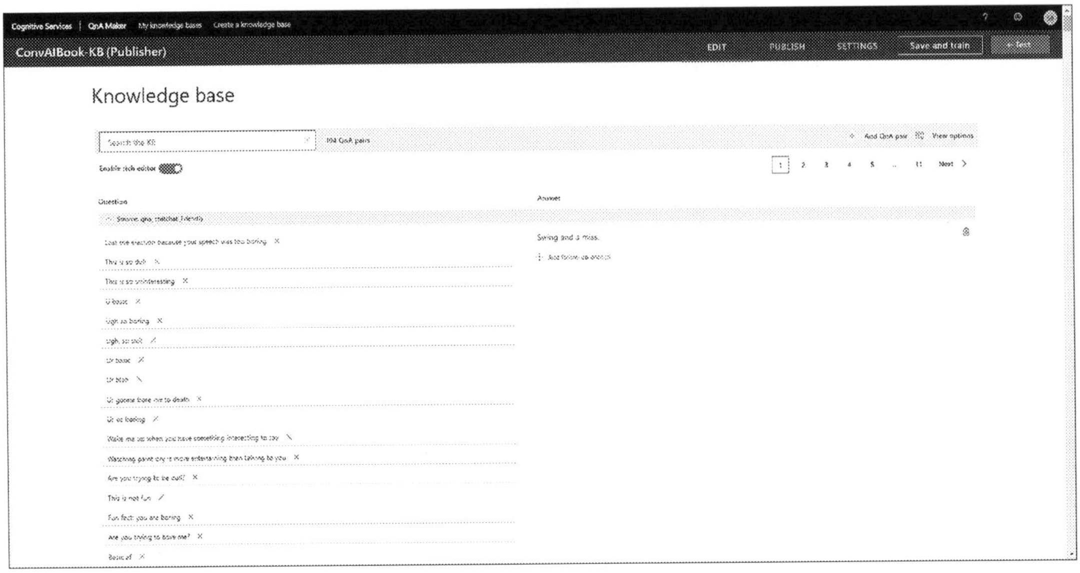

***Figure 3-31.*** *Set up QnA Maker KB 05 – KB after creation*

## Using the CLI

While you can use the QnA Maker portal as mentioned in the previous steps to create a KB, you can also achieve the same result using the CLI. To create a new KB using the Bot Framework CLI, you need at first create a new .qna file, where you can add your QnA pairs in the following format:

```
> # QnA Definitions
### ? Why did Microsoft develop the Bot Framework?
    ' ' '
    We created the Bot Framework to make it easier for developers to build
    and connect great bots to users, wherever they converse, including on
    Microsoft's premier channels.
    ' ' '

### ? What is the v4 SDK?
    ' ' '
    Bot Framework v4 SDK builds on the feedback and learnings from the
    prior Bot Framework SDKs. It introduces the right levels of abstraction
    while enabling rich componentization of the bot building blocks. You
```

```
can start with a simple bot and grow your bot in sophistication using
a modular and extensible framework. You can find FAQ for the SDK on
GitHub.
'''
```

After you have created that .qna file, you will need to convert it to a .json file by executing the following command. This is necessary, as the Bot Framework CLI only accepts valid JSON files for successfully creating new knowledge bases:

```
bf qnamaker:convert -i ".\03_convAIBook-QnA.qna" -o ".\03_convAIBook-QnA.
json"
```

After running this command, you should see an output message stating that it successfully wrote the QnA model to the JSON file you specified. The output of the preceding command should look as follows:

```
{
  "urls": [],
  "qnaList": [
    {
      "id": 0,
      "answer": "We created the Bot Framework to make it easier for
      developers to build and connect great bots to users, wherever they
      converse, including on Microsoft's premier channels.",
      "source": "custom editorial",
      "questions": [
        "Why did Microsoft develop the Bot Framework?"
      ],
      "metadata": []
    },
    {
      "id": 0,
      "answer": "Bot Framework v4 SDK builds on the feedback and learnings
      from the prior Bot Framework SDKs. It introduces the right levels
      of abstraction while enabling rich componentization of the bot
      building blocks. You can start with a simple bot and grow your bot in
      sophistication using a modular and extensible framework. You can find
      FAQ for the SDK on GitHub.",
```

```
      "source": "custom editorial",
      "questions": [
        "What is the v4 SDK?"
      ],
      "metadata": []
    }
  ],
  "files": [],
  "name": ""
}
```

The next step is to get your QnA Maker subscription key from the Azure portal under "Keys and Endpoint" of your Azure QnA service instance you provisioned in the previous step. After you have copied your key, you can then execute the following command in your terminal for creating a new KB (remember to change the name of your KB and the path to the .qna file):

```
bf qnamaker:kb:create --name "ConvAIBook-CLI-KB" --subscriptionKey
"yourQnAMakerSubscriptionKey" --in ".\03_convAIBook-QnA.json"
```

After the command executed successfully, you should see an output message including your new knowledge base ID, as outlined in Figure 3-32.

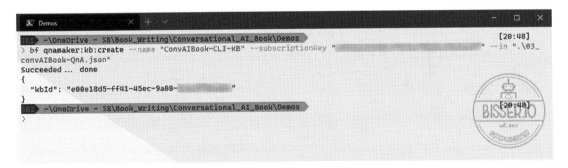

***Figure 3-32.*** *Set up QnA Maker KB using the CLI 01 – output after creation*

If you now switch back to the QnA Maker portal and go to "My knowledge bases" from the top navigation menu, you should see that the new KB "ConvAIBook-CLI-KB" has been created successfully. Additionally the two question and answer pairs you included in the .qna file have also been added to your new knowledge base by the CLI commands, as illustrated in Figure 3-33.

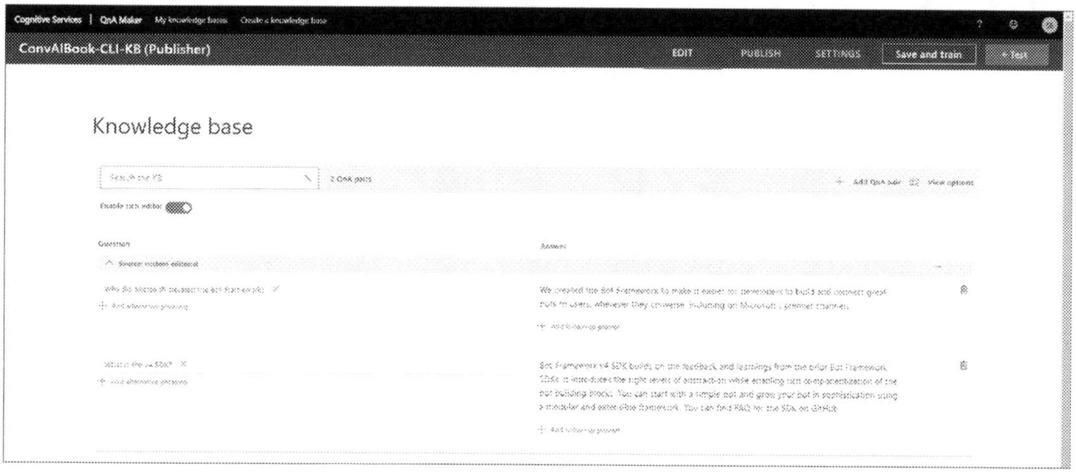

**Figure 3-33.** *Set up QnA Maker KB using the CLI 02 – KB after creation*

# Populating a QnA Maker Knowledge Base

After creating your QnA Maker knowledge base, the process of adding data to your KB begins. For adding new questions and answers, you have the following options which this section of this chapter will cover:

- Adding data manually through the QnA Maker portal

- Adding data using the Bot Framework CLI

- Adding URLs to the KB and letting the QnA Maker service extract QnA pairs automatically

- Adding files directly to the KB through the QnA Maker portal

## Adding Data Manually Through the QnA Maker Portal

This option is especially convenient for QnA content editors, as the QnA Maker portal allows you to add questions and answers in a very easy way. First of all open your knowledge base from `https://qnamaker.ai`, and click the "+ Add QnA pair" button as shown in Figure 3-34.

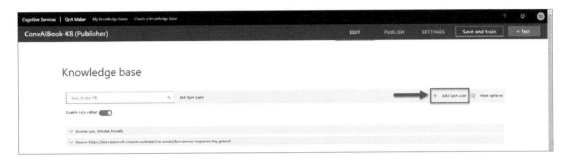

**Figure 3-34.** *Populate KB using the QnA Maker portal 01*

Now you will get a new input box in the "Editorial" source, which is the source where all manually added QnA pairs are aggregated. Within here you can now add a new question in the question column and an appropriate answer in the answer column as outlined in Figure 3-35.

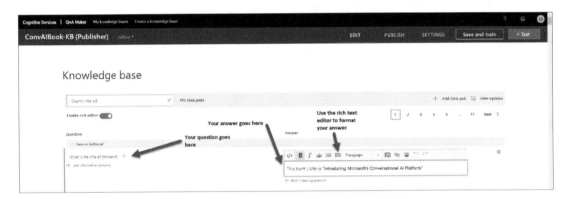

**Figure 3-35.** *Populate KB using the QnA Maker portal 02 – adding QnA pairs*

---

**Note**   The QnA Maker portal offers a simple rich text editor to format your answers and apply some styling like bold, italic, bullet points, or even links, images, and emojis.

---

After adding your first QnA pair, one important thing to do is to add a couple of alternative phrases for your question. It is therefore important, as your users using your conversational application might not always use the exact same question to ask for something. Thus, you need to think about how users might ask certain questions which should lead to the same answer to cover all possible phrasing variations. For instance, the question "What is the title of this book" could also be asked as one of the following example phrases:

- What is the name of that book?

- What's this book named?

- What is this book titled?

- What is the label of this book?

- Could you tell me the name of this copy?

- I want to know the title of this publication please.

- Do you know the name of this work?

All these abovementioned examples cover the question about the name of this book, but each of them is individual. And as QnA Maker is using a pattern-matching algorithm to detect and rank the best possible answer, the best approach for building a sophisticated and intelligent knowledge base is to add as many alternative phrasings as you can think of. This makes sure that your conversational application finds the best-suited answer no matter how users are querying it. So the next step would be to add all of these examples from the preceding text as alternative phrasings to your first question as exposed in Figure 3-36.

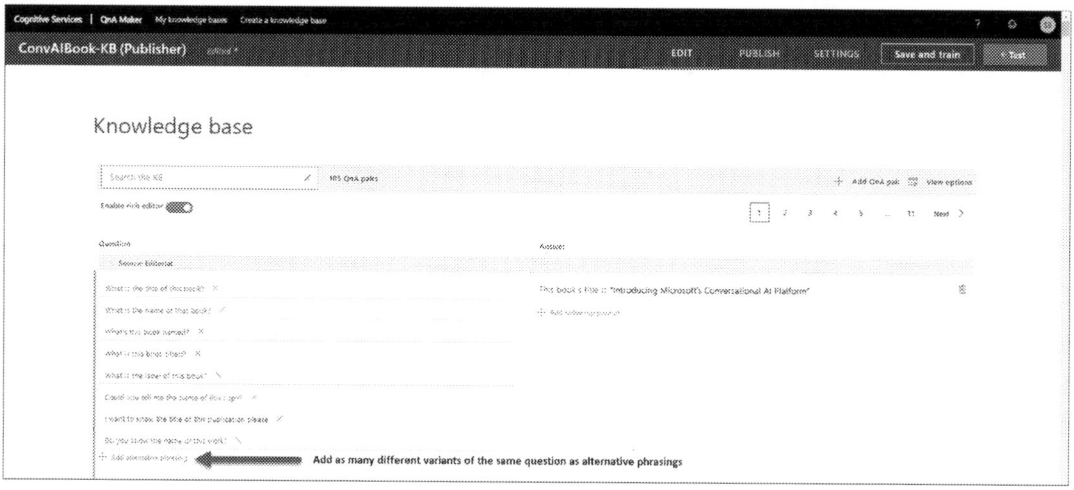

***Figure 3-36.*** *Populate KB using the QnA Maker portal 03 – adding alternative phrasings*

Now you can also link two questions together to build up a context mechanism. Therefore, you can for instance answer the question "What is the title of this book?" and providing a so-called follow-up prompt to provide the user with a matching question on the same topic. This makes it easier for users to navigate through the conversation. As an example, you could add the question "What is the book about?" as a follow-up prompt to the initial question about the book's title. To do this simply click the "+ Add follow-up prompt" in the answer column of a QnA pair as shown in Figure 3-37.

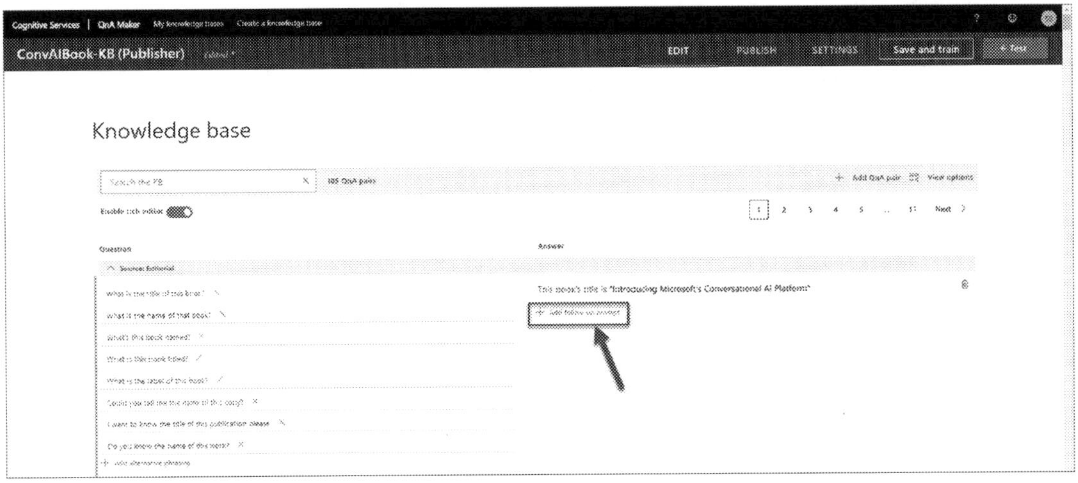

**Figure 3-37.**  *Populate KB using the QnA Maker portal 04 – adding follow-up prompts*

Then a new popup dialog will be opened where you can either enter a new question or pick an existing question from your knowledge base as a follow-up prompt, as outlined in Figure 3-38.

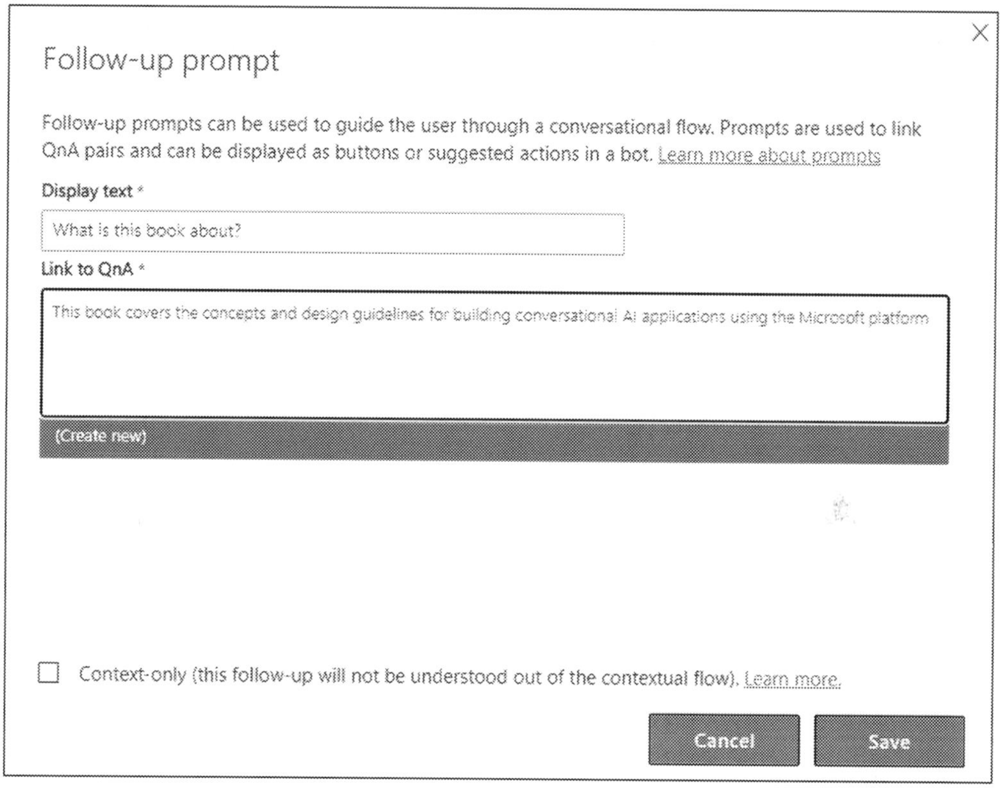

**Figure 3-38.** *Populate KB using the QnA Maker portal 05 – entering follow-up prompt data*

After saving your follow-up prompt, you will notice that a new QnA pair has been added to your knowledge base with the display text you entered as question and the text you entered in the "Link to QnA" as answer. Now you would as well need to add some alternative phrasings to your newly added question to cover a broad range of possible phrasings, like shown in Figure 3-39.

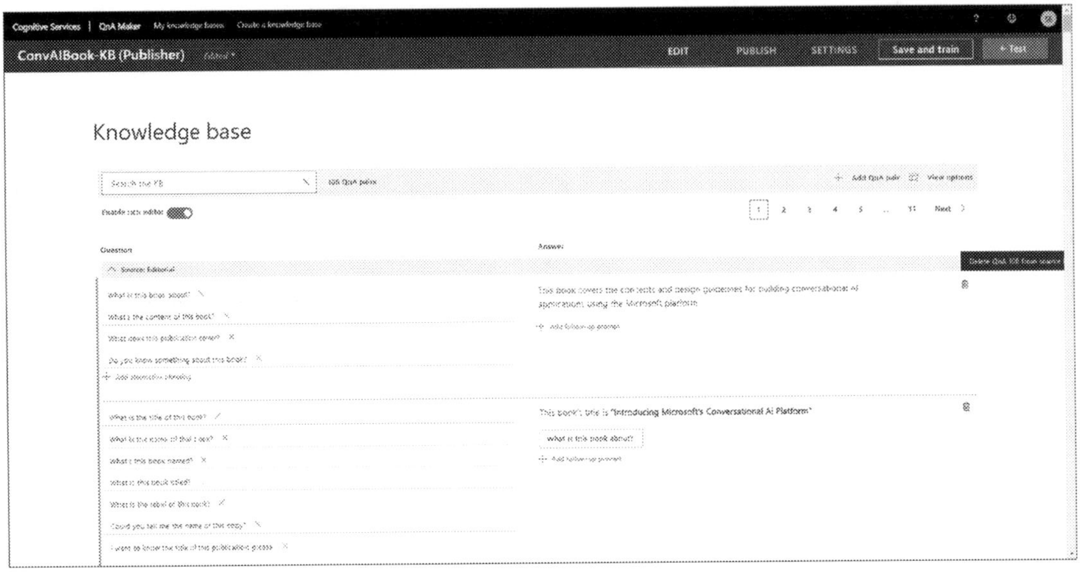

**Figure 3-39.** *Populate KB using the QnA Maker portal 06 – newly added follow-up prompts*

## Adding Data Using the Bot Framework CLI

For adding content to the QnA Maker knowledge base leveraging the Bot Framework CLI, the first thing you need is again a .qna file. This file should include all new QnA pairs, which you want to add to your knowledge base. A very simple example would be the following, where only one QnA pair will be added to the KB later:

```
> # QnA Definitions
### ? Who wrote this book?
    '''

    Stephan Bisser wrote this book as he is passionate about all things
    Conversational AI
    '''
```

After you created the .qna file, you need to use the convert command within the Bot Framework CLI again, to convert the .qna file to a JSON file which can then be used to update the knowledge base. So, you need to execute the following command in your terminal for converting the file to JSON:

```
bf qnamaker:convert -i ".\03_convAIBook-QnA_update.qna" -o ".\03_
convAIBook-QnA_update.json"
```

This command will create a new file *"03_convAIBook-QnA_update.json"* which will eventually look like this:

```
{
  "urls": [],
  "qnaList": [
    {
      "id": 0,
      "answer": "Stephan Bisser wrote this book as he is passionate about
      all things Conversational AI",
      "source": "custom editorial",
      "questions": [
        "Who wrote this book?"
      ],
      "metadata": []
    }
  ],
  "files": [],
  "name": ""
}
```

But in order to execute the next command for updating the knowledge base accordingly, you need to change the file manually so that it looks like this:

```
{
  "urls": [],
  "add":{
    "qnaList": [
      {
        "id": 0,
        "answer": "Stephan Bisser wrote this book as he is passionate about
        all things Conversational AI",
        "source": "Editorial",
        "questions": [
          "Who wrote this book?"
        ],
```

```
        "metadata": []
      }
   ],
   "files": [],
   "name": ""
  }
}
```

In the preceding example of the *"03_convAIBook-QnA_update.json"* file, you see that the *"qnaList"* has been surrounded by an *"add"* object to indicate that these QnA pairs should be added to the knowledge base. You could also use *"delete"* for deleting specific QnA pairs or *"update"* for updating QnA pairs as well. Additionally, the source of the "qnaList" object has been changed from "custom editorial" to *"editorial"* as the default source after creating a new KB which is called *"editorial"* as mentioned earlier. This makes sure that the QnA pairs will be added to the correct source in the knowledge base later. After these changes have been done, the next step is to execute the following command to update the knowledge base. Do not forget to insert your knowledge base ID and your QnA Maker subscription key in there before executing the command in the terminal:

```
bf qnamaker:kb:update --kbId "yourQnAMakerKBId" --subscriptionKey
"yourQnAMakerSubscriptionKey" --in ".\03_convAIBook-QnA_update.json"
```

The output of this command will be an information about this executed operation, as shown in Figure 3-40.

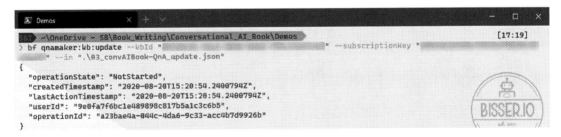

***Figure 3-40.*** *Populate KB using the Bot Framework CLI 01 – operation status*

To check whether the operation is already finished, execute the following command in the terminal:

```
bf qnamaker:operationdetails:get --operationId "yourQnAMakerOperationId"
--subscriptionKey "yourQnAMakerSubscriptionKey"
```

If the operation has been executed successfully, the *"operationState"* value of the output should state *"Succeeded,"* as illustrated in Figure 3-41.

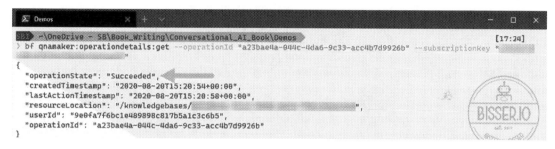

***Figure 3-41.*** *Populate KB using the Bot Framework CLI 02 – operation status check*

If you now go back to your knowledge base in the QnA Maker portal, you should see that the QnA pair from your .qna file has been added to your knowledge base in the "Editorial" source, as exposed in Figure 3-42.

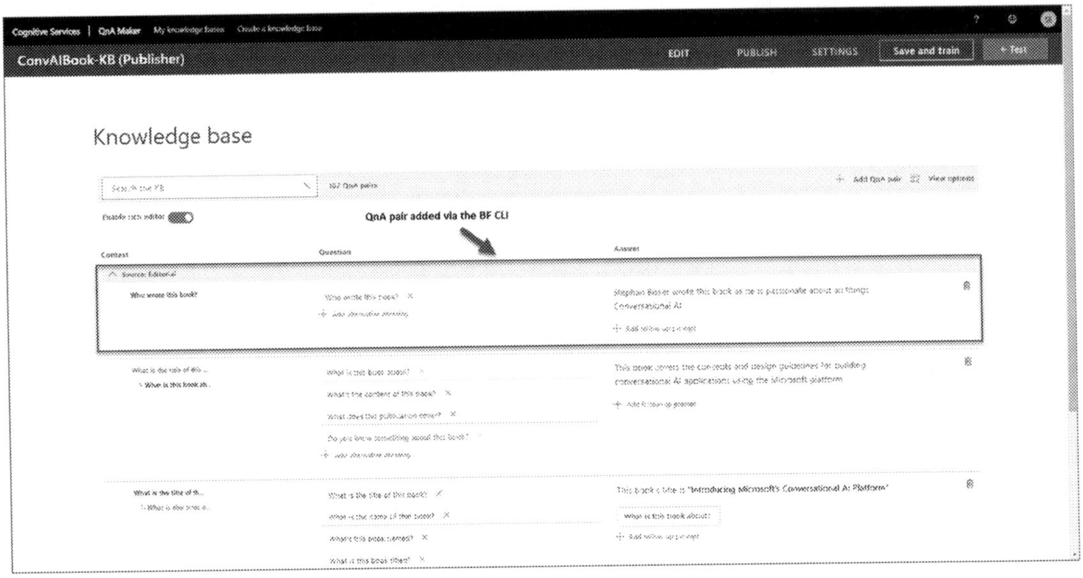

**Figure 3-42.** *Populate KB using the Bot Framework CLI 03 – newly added QnA pair*

## Adding URLs to the KB

In contrast to adding QnA pairs manually to a QnA Maker knowledge base, you can also let QnA Maker extract content which is already available on a website. Therefore, you do not need to enter the data manually or via the CLI but import content from an URL. This can be done by opening your knowledge base in the QnA Maker portal and switching to the *"Settings"* page. In the *"Manage knowledge base"* section, you can add URLs, like this one `https://docs.microsoft.com/en-us/azure/cognitive-services/QnAMaker/troubleshooting` as shown in Figure 3-43.

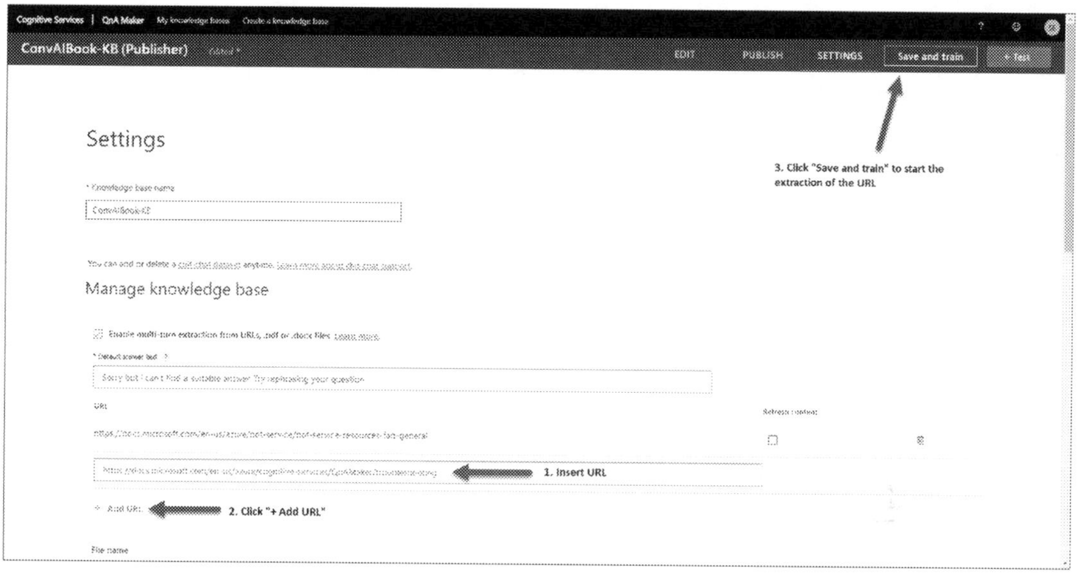

**Figure 3-43.** *Adding URLs to the QnA Maker KB 01*

After adding the URL and clicking "Save and train" to start the extraction process, the content of the website should soon be added to your knowledge base, whereas the headings within the website (basically the <h_n></h_n> html tags) will be extracted and added as questions to your knowledge base and the paragraphs (meaning the <p></p> html tags) will be treated as the answers to those questions, as outlined in Figure 3-44.

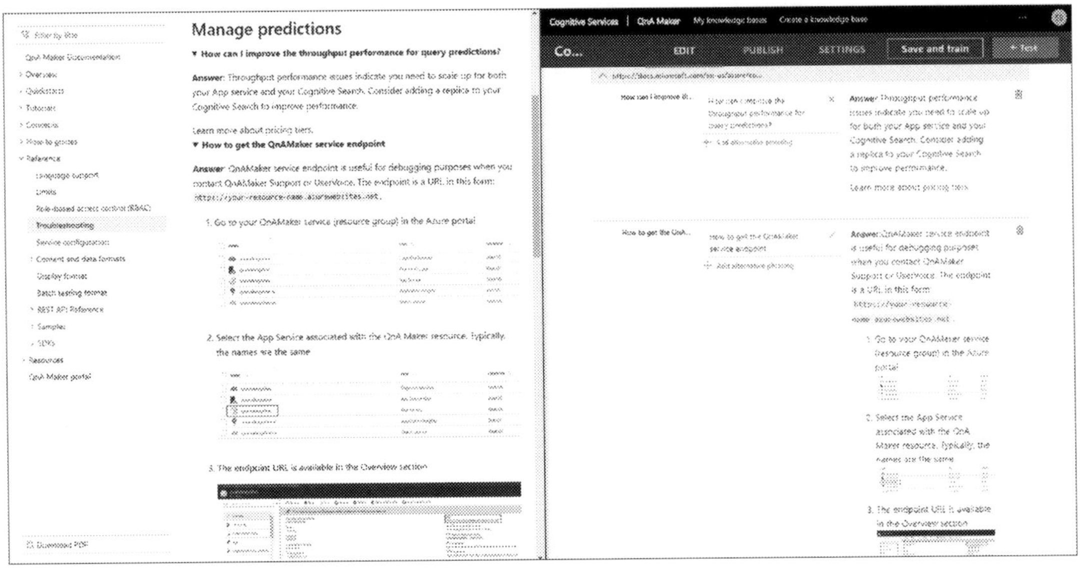

**Figure 3-44.** *Adding URLs to the QnA Maker KB 02 – comparison (left original website, right KB)*

If the website's content has been updated, you can also update your knowledge base. If you need to update the content of your KB which has been extracted from URLs, you need to go to the settings page of your KB in the QnA Maker portal and select the "Refresh content" check box followed by "Save and train" to start the update process for the selected URLs, as illustrated in Figure 3-45.

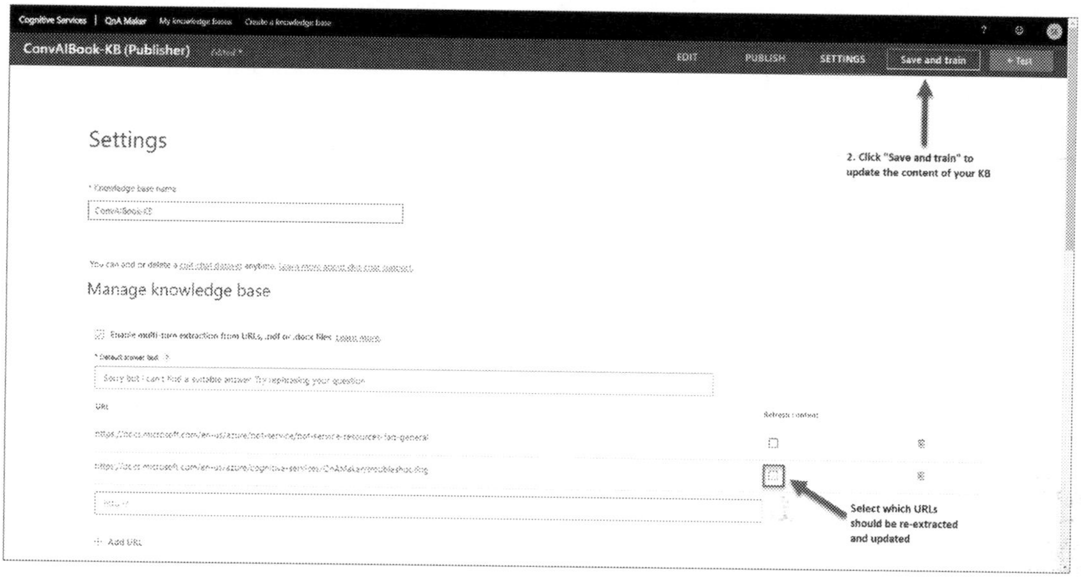

**Figure 3-45.** *Adding URLs to the QnA Maker KB 03 – update content extracted from URLs*

---

**Note** The content of the website URL(s) you add to your knowledge base will not be updated automatically if the website's content has been changed. To update your KB, you will need to trigger the update manually or via the QnA Maker API.

---

## Adding Files Directly to the KB

In addition to adding URLs to a QnA Maker knowledge base, you can also upload files directly. This is especially helpful if you already built FAQ files, which are not part of a public-facing website. Table 3-2 includes all supported file or data types which can be imported into a knowledge base directly. All these file and document types can be used to populate your knowledge base without the need of manually adding QnA pairs directly into the KB. However, the files should be built in a FAQ-like style. This means it should not be a file with paragraphs which span over a couple of pages, but it should be constructed in a way that the QnA Maker service can extract the content in a way that QnA pairs can be parsed in a reasonable fashion.

113

***Table 3-2.***  *QnA Maker-supported file types (Microsoft, 2020)*

| Source Type | Content Type |
| --- | --- |
| **PDF/Word (.pdf/.docx)** | FAQs, product manual, brochures, paper, flyer policy, support guide, structured QnA, etc. |
| **Excel (.xlsx)** | Structured QnA file (including RTF, HTML support) |
| **TXT/TSV (.txt/.tsv)** | Structured QnA file |

> **Note**    For detailed examples of each supported data type in QnA Maker, conduct `https://docs.microsoft.com/en-us/azure/cognitive-services/ qnamaker/concepts/content-types`.

To give an example, take a look at the Word file, which can be downloaded from `http://bit.ly/ConvAIBook-structured-qna`. This Word file basically contains multiple first-level headlines followed by paragraphs. The headlines are mainly framed as questions, whereas the paragraphs are the suitable answers to those questions.

In order to add a file from your machine to the knowledge base, go to the "Settings" page in your KB, and select "+ Add file" under the "Manage knowledge base" section, as illustrated in Figure 3-46. After clicking the button, an upload dialog opens, where you can select the right file and upload it your knowledge base.

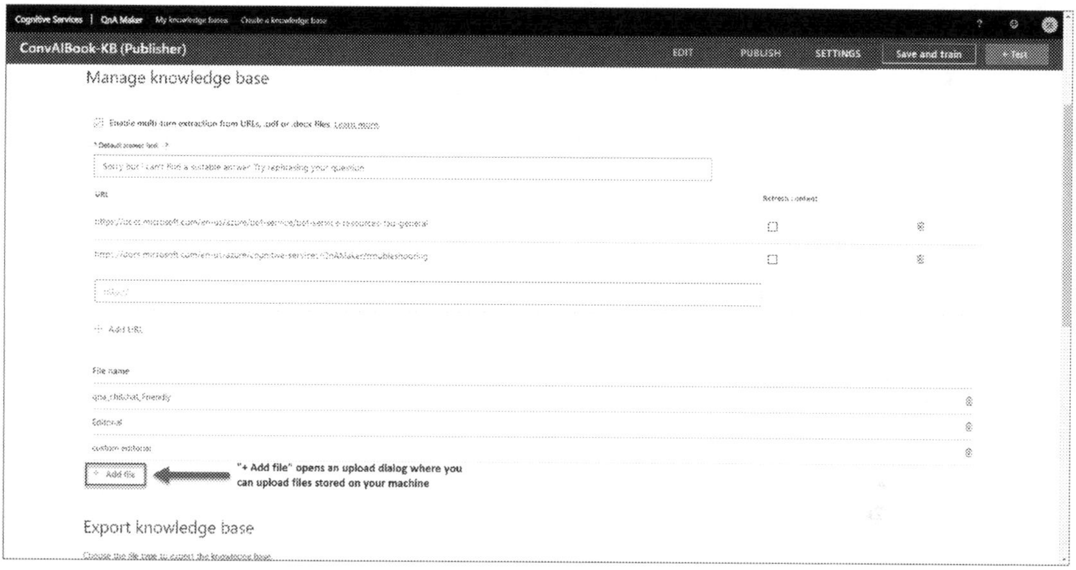

**Figure 3-46.** *Adding files to the QnA Maker KB 01*

Now that the file has been uploaded, you need to save and train your knowledge base again, to let the QnA Maker service extract the contents of the file. If you then switch back to the "Edit" page of the KB, you will see that it contains all QnA pairs, which are part of the file you uploaded, as outlined in Figure 3-47.

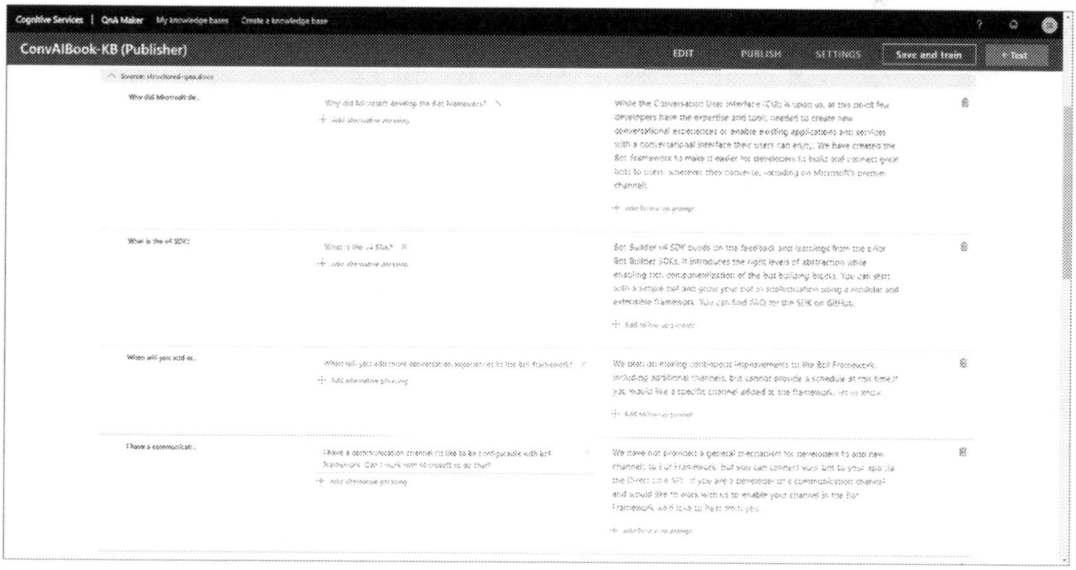

**Figure 3-47.** *Adding files to the QnA Maker KB 02*

# Testing a KB

One advantage of using the QnA Maker portal is that it offers a testing experience. This is extremely helpful and should be used when you update your knowledge base with new QnA pairs. Testing a knowledge base is crucial to the success of your conversational application, as you can detect and fix low confidence or incorrect query results within your KB before your users will be impacted by those issues.

To test a knowledge base, go to your KB in the QnA Maker portal and select the "← Test" button from the upper right corner. This will bring up the interactive testing panel, where you can enter sample queries, which you think users might ask. Remember to use different query data than the data which is stored in your KB, as users might not always use the exact same phrasing as the questions which are stored in your knowledge base. For instance, in your knowledge base, you should have the question "What is this book about?" included. For testing this specific question, think of a different way of phrasing this question, for example, "What is the gist of that publication?" Now you can use the testing panel and insert this phrase to see if the answer is correct or not. As Figure 3-48 points out, the answer which is returned by the service is not quite the one you would want it to be, as it responds with the answer about the book's title, not the content. As you can also see from that figure, the confidence score which is 49.85 is quite low, so this question needs some improvements.

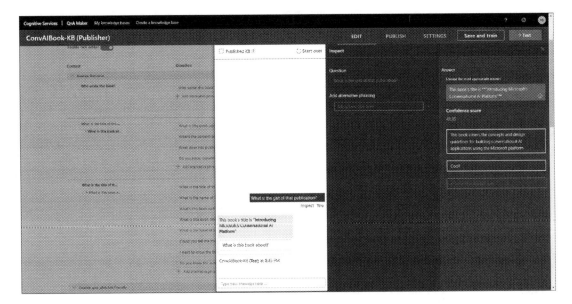

***Figure 3-48.***  *Testing a QnA Maker KB 01*

Another example which you could use for testing the same question would be "What is the essence of this book?" If you enter this question into the testing panel, you should see that the service responds with the correct answer about the book's content, although the confidence score with 68.69 is still low, as illustrated in Figure 3-49.

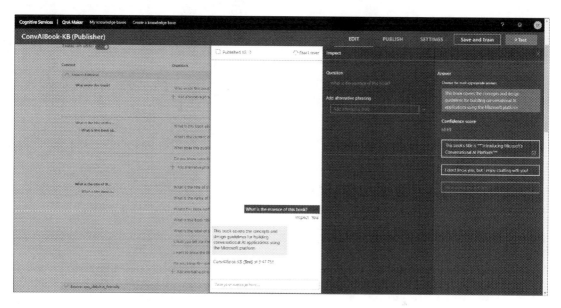

***Figure 3-49.*** *Testing a QnA Maker KB 02*

In order to fix these two issues, you could now go ahead and add "What is the gist of that publication?" and "What is the essence of this book?" as alternative phrasings to the question "What is this book about?" to tell the service that these kinds of questions are covering the topic of the book's content not the title. After saving and training your knowledge base, you can go ahead and test again, using the query "What is the essence of that publication?" for example, and you will notice that the answer is given back correctly and the confidence score with 94.17 is also acceptable, even though the query is not exactly part of a QnA pair, but is slightly different worded, as outlined in Figure 3-50.

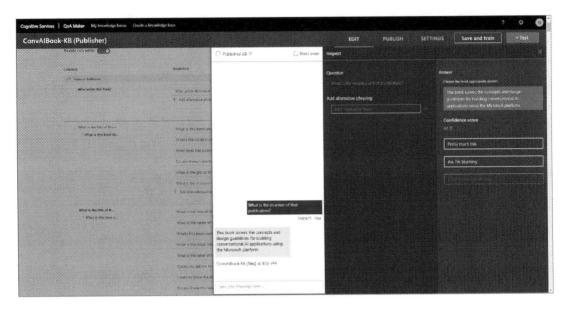

***Figure 3-50.*** *Testing a QnA Maker KB 03*

---

**Note**   Testing a knowledge base is an essential process and should be done extensively, to fix low confidence scores and detect incorrect query results.

---

## Publishing a KB

The last step within your knowledge base life cycle before users can access the content of your KB through a conversational application like a chatbot is to publish the KB. Publishing a knowledge base basically consists of two things:

- The latest version of your KB gets stored in a dedicated index within the Azure Search service.

- An endpoint is made available which can be called from your conversational application.

To publish a knowledge base, go to the *"Publish"* page within your QnA Maker portal, and click the "Publish" button, as highlighted in Figure 3-51.

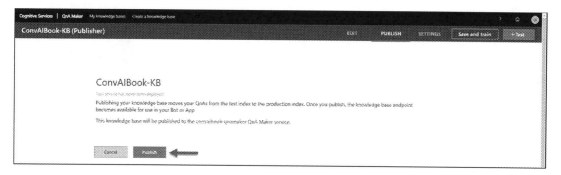

**Figure 3-51.** *Publishing a QnA Maker KB 01*

After the knowledge base has been published, you will see all endpoint details, which you will need later, when calling your QnA Maker service from a chatbot for instance, as outlined in Figure 3-52:

- **Knowledge base ID**: The ID of your knowledge base is included in the POST request URL sample after "/knowledgebases/" and is needed to determine the correct KB.

- **Host**: The host is the endpoint which you need to call from your conversational application when querying your KB.

- **Authorization**: The EndpointKey is used to authorize requests against your knowledge base to prevent unauthorized applications from accessing your KB's content.

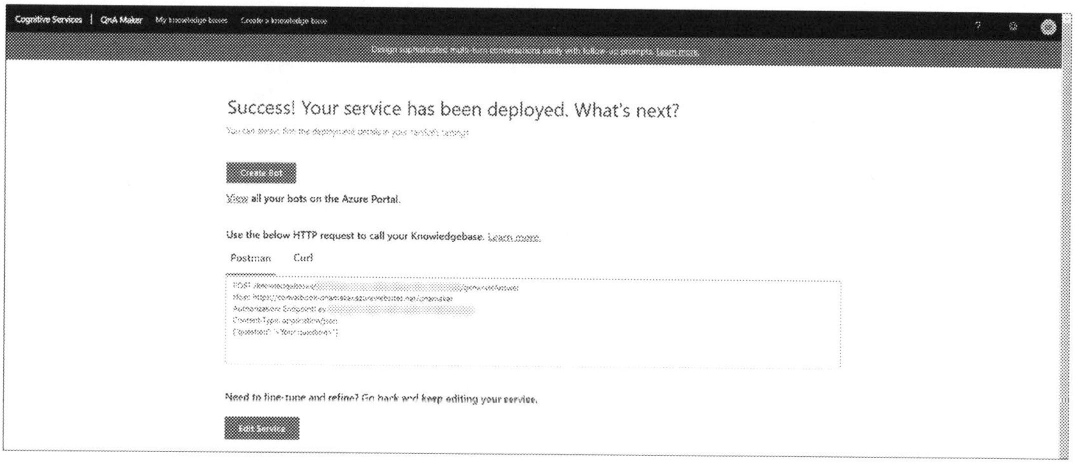

**Figure 3-52.** *Publishing a QnA Maker KB 02*

Of course, you could also use the Bot Framework CLI to publish your knowledge base executing the following command in a terminal session, if you do not want to publish the KB via the QnA Maker portal or if you want to automate the publishing of your knowledge base at some point:

```
bf qnamaker:kb:publish --kbId "yourQnAMakerKBId" --subscriptionKey
"yourQnAMakerSubscriptionKey"
```

# Text Analytics

Another service within the language category, which is quite often used for conversational applications, is the Text Analytics API. This is a service covering mainly the following four use cases:

- Sentiment analysis

- Key phrase extraction

- Named entity recognition

- Language detection

Many of these use cases are somehow beneficial when developing a chatbot or any other conversational application, as they enhance the natural language processing features. As it is a ready-to-use API, there is no need of customizing the components within that service. Therefore, the next section will cover all four of these use cases in detail to explain the concepts behind them; however, the practical implementation will follow in later chapters.

## Sentiment Analysis

Imagine you build a chatbot which is included in your customer service process, to act as the first gate where user inquiries are processed in a first place. In many cases, your chatbot will be able to answer a limited set of questions, mostly frequently asked ones, but in other cases, your chatbot will not be able to answer a specific user question sufficiently. This is a crucial point, as you do not want to lose your users, when the chatbot is not capable of answering a question, so you need to think about a human handoff of some sort. Human handoff basically describes the process of handing off the conversation from the chatbot to a person, which can help the user, as the chatbot cannot answer the user's question anymore. But within this process, you certainly need to decide when this human handoff should occur, as it does not make sense to hand off each time the chatbot is not able to answer a specific question. This is basically the perfect situation where sentiment detection could help you find the right time to perform the human handoff.

The sentiment analysis feature within the Text Analytics API provides you with a sentiment score between 0 and 1 for a given text. The aim of this feature is to supply you with the knowledge about the person's sentiment with which you will be able to decide if the person is either in a positive or negative mood currently. A simple example would be the sentence "I was really pleased with the nice service in the restaurant yesterday and I would definitely come back." If you call the Text Analytics API's sentiment analysis feature with that example sentence, you will receive a sentiment score result of 0.90463, meaning the sentence is quite positive, as outlined in Figure 3-53.

```
Transfer-Encoding: chunked
csp-billing-usage: CognitiveServices.TextAnalytics.BatchScoring=1
x-envoy-upstream-service-time: 11
apim-request-id: 4edc3614-ccd7-4083-af9f-1f4d843164cf
Strict-Transport-Security: max-age=31536000; includeSubDomains; preload
x-content-type-options: nosniff
Date: Fri, 21 Aug 2020 11:28:12 GMT
Content-Type: application/json; charset=utf-8

{
  "documents": [{
    "id": "1",
    "score": 0.90463638305664063
  }],
  "errors": []
}
```

***Figure 3-53.*** *Text Analytics API sentiment analysis feature 01 – positive example*

However, if you would use the sentence "I was surprised that service in the restaurant yesterday was really awful which is why I would never ever come back again" and call the Text Analytics API with that sentence, you will see that the sentiment score with 0.1873 is pretty low as illustrated in Figure 3-54. This means that the sentence is rather negative in that case.

```
Transfer-Encoding: chunked
csp-billing-usage: CognitiveServices.TextAnalytics.BatchScoring=1
x-envoy-upstream-service-time: 10
apim-request-id: 3ce26b48-d78f-42ac-bf5f-7fa80277e326
Strict-Transport-Security: max-age=31536000; includeSubDomains; preload
x-content-type-options: nosniff
Date: Fri, 21 Aug 2020 11:27:45 GMT
Content-Type: application/json; charset=utf-8

{
  "documents": [{
    "id": "1",
    "score": 0.18730971217155457
  }],
  "errors": []
}
```

*Figure 3-54. Text Analytics API sentiment analysis feature 02 – negative example*

By using the sentiment analysis feature of the Text Analytics API, you can easily detect the user's mood in deciding if the chatbot should directly hand off the conversation to a person, if the user is in a negative mood. This can have positive effects on the user's situation as it might somehow act as a deescalation scenario, because the person most probably answers the user's questions more efficiently. This in turn leads to the fact that the user might come back again which has a positive impact on your service offerings.

## Key Phrase Extraction

Key phrase extraction is another feature within the Text Analytics API, which could bring some advantages to your conversational application. To stick with the example mentioned earlier about a chatbot which is integrated into your customer service solution, imagine that users would send rather long messages to your chatbot, like "I was really pleased with the nice service in the restaurant yesterday and I would definitely come back. However, I did not like the fact that the soup was almost cold whereas the main dish was too hot to eat. But the ice cream dessert was perfect!". This message is rather long, so the chatbot would need to analyze not only one sentence but more. This could be a challenging task, especially when human handoff should be accomplished, as there could be more experts involved in here.

Therefore, the ability to detect certain key phrases within a user inquiry could help detecting the important parts within a sentence and then act accordingly. In Figure 3-55 for instance, you see the result of the abovementioned example phrase. The Text Analytics key phrase extraction API has detected the following key phrases within that sentence:

- Main dish

- Soup

- Ice cream dessert

- Nice service

- Restaurant

- Fact

In these key phrases, you see that, for instance, nice service has been extracted. Thus, you could conclude that the person who wrote that message wants to give feedback that the service was nice and the restaurant visit was pleasant.

```
Transfer-Encoding: chunked
csp-billing-usage: CognitiveServices.TextAnalytics.BatchScoring=1
x-envoy-upstream-service-time: 20
apim-request-id: 9e71461d-f127-4d0b-816f-361b2aaf7d51
Strict-Transport-Security: max-age=31536000; includeSubDomains; preload
x-content-type-options: nosniff
Date: Sun, 23 Aug 2020 09:16:00 GMT
Content-Type: application/json; charset=utf-8

{
  "documents": [{
    "id": "1",
    "keyPhrases": ["main dish", "soup", "ice cream dessert", "nice service", "restaurant", "fact"],
    "warnings": []
  }],
  "errors": [],
  "modelVersion": "2020-07-01"
}
```

**Figure 3-55.** *Text Analytics API key phrase extraction feature 01*

Combining the key phrase extraction feature and the sentiment analysis service, the human handoff part within a chatbot could be improved heavily. The reason for this is that, first of all, you could detect the sentiment of the user's input to see if the user is in a good or bad mood currently. Depending on that, you could either let the chatbot finish the conversation if the sentiment score is positive or hand off the conversation to an agent if the sentiment score is negative. If you then extract the key phrases of the user's messages and hand them over to the agent for preview, the agent might know the general topic more quickly, as if he or she would need to read the complete chat transcript. So, the agent can take over the conversation faster and help the user in a timelier fashion.

## Named Entity Recognition

Named entity recognition accepts unstructured text and gives back entities which the API recognizes. These could be either person names, locations, events, products, or organizations, which are well-known. An example for that would be the sentence "The restaurant next to the Space Needle in Seattle where I met Bill Gates was really good!" As Figure 3-56 outlines, the words "Space Needle," "Seattle," and "Bill Gates" have been recognized, and the API even marks the corresponding category like location or person for the recognized entities.

```
Transfer-Encoding: chunked
csp-billing-usage: CognitiveServices.TextAnalytics.BatchScoring=1
x-envoy-upstream-service-time: 92
apim-request-id: c04545ab-4c28-4ada-9a7f-3b9e3380f2a8
Strict-Transport-Security: max-age=31536000; includeSubDomains; preload
x-content-type-options: nosniff
Date: Sun, 23 Aug 2020 09:47:05 GMT
Content-Type: application/json; charset=utf-8

{
  "documents": [{
    "id": "1",
    "entities": [{
      "text": "Space Needle",
      "category": "Location",
      "offset": 27,
      "length": 12,
      "confidenceScore": 0.36
    }, {
      "text": "Seattle",
      "category": "Location",
      "subcategory": "GPE",
      "offset": 43,
      "length": 7,
      "confidenceScore": 0.66
    }, {
      "text": "Bill Gates",
      "category": "Person",
      "offset": 63,
      "length": 10,
      "confidenceScore": 0.84
    }],
    "warnings": []
  }],
  "errors": [],
  "modelVersion": "2020-04-01"
}
```

***Figure 3-56.*** *Text Analytics API named entity recognition feature example*

Another feature within that API is the entity linking service. This basically describes the ability to identify certain entities and their identities within a given text. These recognized entities are basically served from a knowledge base, where all these words or phrases along with links to Wikipedia articles, where they are described, are stored. As an example, the entity linking feature could detect the word "Nike" within a given sentence and could distinguish if the sports brand, the Greek goddess, or the US Army's missile project is meant. As an example, Figure 3-57 shows the output which has been detected from the sentence "The T-Shirt from Nike is really comfortable and I enjoy wearing it."

```
Transfer-Encoding: chunked
csp-billing-usage: CognitiveServices.TextAnalytics.BatchScoring=1
x-envoy-upstream-service-time: 28
apim-request-id: 44975c9d-e1bb-4581-858a-2c6321ac3229
Strict-Transport-Security: max-age=31536000; includeSubDomains; preload
x-content-type-options: nosniff
Date: Sun, 23 Aug 2020 09:42:56 GMT
Content-Type: application/json; charset=utf-8

{
  "documents": [{
    "id": "1",
    "entities": [{
      "name": "T-shirt",
      "matches": [{
        "text": "T-Shirt",
        "offset": 4,
        "length": 7,
        "confidenceScore": 0.22
      }],
      "language": "en",
      "id": "T-shirt",
      "url": "https://en.wikipedia.org/wiki/T-shirt",
      "dataSource": "Wikipedia"
    }, {
      "name": "Nike, Inc.",
      "matches": [{
        "text": "Nike",
        "offset": 17,
        "length": 4,
        "confidenceScore": 0.23
      }],
      "language": "en",
      "id": "Nike, Inc.",
      "url": "https://en.wikipedia.org/wiki/Nike,_Inc.",
      "dataSource": "Wikipedia"
    }],
    "warnings": []
  }],
  "errors": [],
  "modelVersion": "2020-02-01"
}
```

**Figure 3-57.**  *Text Analytics API named entity recognition feature – entity linking example*

## Language Detection

The fourth component within the Text Analytics API is the language detection feature. This service takes unstructured text as an input and responds with the detected language of that given text. This could be helpful in situations where you build a chatbot for an online store, for example, where you expect to get visitors from all around the world. In such situations it is helpful to first detect the language of a user within the conversation before taking the input text to either LUIS or a QnA Maker knowledge base, as a user question in Chinese might not result in the correct answer, when querying an English knowledge base with it. Therefore, this API endpoint can be used to detect the language of one or more sentences, as shown in Figure 3-58, where the input is consisting of the following two phrases:

- "The restaurant next to the Space Needle in Seattle where I met Bill Gates was really good!" – English

- "Das Restaurant neben der Space Needle in Seattle in dem ich Bill Gates getroffen habe war exzellent!" – German

```
Transfer-Encoding: chunked
csp-billing-usage: CognitiveServices.TextAnalytics.BatchScoring=2
x-envoy-upstream-service-time: 33
apim-request-id: 17ea6072-3de6-4b3d-859b-46361f5b4d87
Strict-Transport-Security: max-age=31536000; includeSubDomains; preload
x-content-type-options: nosniff
Date: Sun, 23 Aug 2020 09:54:59 GMT
Content-Type: application/json; charset=utf-8

{
  "documents": [{
    "id": "1",
    "detectedLanguage": {
      "name": "English",
      "iso6391Name": "en",
      "confidenceScore": 1.0
    },
    "warnings": []
  }, {
    "id": "2",
    "detectedLanguage": {
      "name": "German",
      "iso6391Name": "de",
      "confidenceScore": 1.0
    },
    "warnings": []
  }],
  "errors": [],
  "modelVersion": "2020-07-01"
}
```

*Figure 3-58. Text Analytics API language detection feature example*

# Translator

As mentioned in Chapter 1, translation is an essential component within a chatbot use case, as it eases the process of data management. By introducing translation to your chatbot solution, you do not need to create knowledge bases, language models, and other crucial services for each language individually, as you can basically keep all of your "back-end" components in one or two languages and translate every input from another language into one of those languages before further processing it. The Translator service within Azure Cognitive Services offers the following features, which can be used in a conversational application scenario:

- Translate text.

- Transliterate text.

- Detect language.

- Look up word translations.

- Determine sentence length.

## Translate

The most used feature is, of course, the translate text feature. This feature lets you translate sentences from 1 language into more than 70 languages. The use of the translate feature is rather easy. As it is an API, you only need to create an HTTP POST call in the simplest scenario, provide your Azure subscription key for the Translator resource, choose which language the text should be translated to, and add the text which should be translated, as shown in the example curl request here:

```
curl -X POST "https://api.cognitive.microsofttranslator.com/
translate?api-version=3.0&from=en&to=de" -H "Ocp-Apim-Subscription-Key:
yourTranslatorKey" -H "Content-Type: application/json; charset=UTF-8" -d
"[{'Text':'Hello, what is your name?'}]"
```

The result will be the translation of the given text in the language you picked, as seen here:

```
[{
    "translations":[
        {
            "text":"Hallo, wie heißt du?",
            "to":"de"
        }
    ]
}]
```

Additionally, you can also translate text into multiple languages at once as well, by adding all languages to your HTTP request like in this example:

```
curl -X POST "https://api.cognitive.microsofttranslator.com/translate?api-version=3.0&from=en&to=de&to=it&to=ja&to=th" -H "Ocp-Apim-Subscription-Key: yourTranslatorKey" -H "Content-Type: application/json; charset=UTF-8" -d "[{'Text':Hi how are you?'}]"
```

As a result, you will get back the translated sentences in all chosen languages:

```
[{
    "translations": [
        {
            "text":"Hallo, wie geht es dir?",
            "to":"de"
        },
        {
            "text":"Ciao come stai?",
            "to":"it"
        },
        {
            "text":"やあ、元気。",
            "to":"ja"
        },
```

```
        {
                "text":"ไง เธอเป็นไงบ้าง",
                "to":"th"
        }
    ]
}]
```

The translation feature can be very useful in scenarios where you build a multilingual chatbot, but do not want to build up LUIS applications or QnA Maker knowledge bases for each language you want to support. In such a scenario, you can pick the most important languages and design your language models and knowledge bases specifically for these languages, which are the most commonly used ones and translate user messages from other languages in one of those to get the best possible result with the least amount of effort.

---

**Note**   For a detailed language support of the Translator API, please conduct
`https://docs.microsoft.com/en-us/azure/cognitive-services/`
`translator/language-support`.

---

## Detect Language

When building a multilingual chatbot, the ability to detect the user's language before applying natural language processing functionalities like language understanding or key phrase extraction is crucial. Therefore, the Translator API offers a feature for detecting the language of a given sentence. This can be extremely helpful in scenarios where you have multiple LUIS applications and QnA Maker KBs in multiple languages deployed, and you need to determine which language to use for a given user input. Here you can make use of the detect language feature to first of all get the correct language of the user input and then query LUIS or QnA Maker to process the user input accordingly. The following HTTP request demonstrates a very basic example for using this feature with the sentence "What language is this text written in?"

```
curl -X POST "https://api.cognitive.microsofttranslator.com/detect?api-version=3.0" -H "Ocp-Apim-Subscription-Key: yourTranslatorKey" -H "Content-Type: application/json" -d "[{'Text':'What language is this text written in?'}]"
```

The result of the preceding HTTP POST call is that it detected English as the language of the preceding sentence. As alternatives Filipino and Irish have also been identified:

```
[{
    "language":"en",
    "score":1.0,
    "isTranslationSupported":true,
    "isTransliterationSupported":false,
    "alternatives":
    [
        {
                "language":"fil",
                "score":1.0,
                "isTranslationSupported":true,
                "isTransliterationSupported":false
        },
        {
                "language":"ga",
                "score":1.0,
                "isTranslationSupported":true,
                "isTransliterationSupported":false
        }
    ]
}]
```

## Dictionary Lookup

In some scenarios it could be helpful to get alternative translations for specific words or phrases. Translator offers a feature for getting such alternate translations along with sample use cases for these translations. For instance, the word "key" should be translated from English to German. By using the dictionary lookup feature, like in the example here, you would get all possible translations back from the API, which allows you to pick the most appropriate one within the current context:

```
curl -X POST "https://api.cognitive.microsofttranslator.com/dictionary/
lookup?api-version=3.0&from=en&to=es" -H "Ocp-Apim-Subscription-Key:
yourTranslatorKey" -H "Content-Type: application/json" -d "[{'Text':'key'}]"
```

The result of the preceding sample request would be the following:

```
[
  {
    "normalizedSource": "key",
    "displaySource": "key",
    "translations": [
      {
        "normalizedTarget": "taste",
        "displayTarget": "taste",
        "posTag": "NOUN",
        "confidence": 0.7575,
        "prefixWord": "",
        "backTranslations": [
          {
            "normalizedText": "key",
            "displayText": "key",
            "numExamples": 0,
            "frequencyCount": 9393
          }
        ]
      },
      {
        "normalizedTarget": "wichtigste",
        "displayTarget": "wichtigste",
        "posTag": "ADJ",
        "confidence": 0.1358,
        "prefixWord": "",
        "backTranslations": [
          {
            "normalizedText": "key",
            "displayText": "key",
            "numExamples": 0,
            "frequencyCount": 1684
          }
        ]
      },
```

```json
{
  "normalizedTarget": "eingeben",
  "displayTarget": "eingeben",
  "posTag": "VERB",
  "confidence": 0.0612,
  "prefixWord": "",
  "backTranslations": [
    {
      "normalizedText": "enter",
      "displayText": "enter",
      "numExamples": 0,
      "frequencyCount": 13370
    },
    {
      "normalizedText": "key",
      "displayText": "key",
      "numExamples": 0,
      "frequencyCount": 759
    },
    {
      "normalizedText": "input",
      "displayText": "input",
      "numExamples": 0,
      "frequencyCount": 732
    }
  ]
},
{
  "normalizedTarget": "tonart",
  "displayTarget": "tonart",
  "posTag": "NOUN",
  "confidence": 0.0455,
  "prefixWord": "",
```

```
      "backTranslations": [
        {
          "normalizedText": "key",
          "displayText": "key",
          "numExamples": 1,
          "frequencyCount": 564
        }
      ]
    }
  ]
  }
]
```

With each of these word lookups, you can now use the Translator to retrieve dictionary examples. This feature demonstrates how these translations are used in context, which gives you a clearer understanding of which of these translations fits best. To retrieve the dictionary examples, you could use the following HTTP request if you would like to retrieve sample phrases with the German word *"wichtigste"* which would be translated to *"most important"* in English:

```
curl -X POST "https://api.cognitive.microsofttranslator.com/dictionary/
examples?api-version=3.0&from=en&to=de" -H "Ocp-Apim-Subscription-Key:
yourTranslatorKey" -H "Content-Type: application/json" -d "[{'Text':'key',
'Translation':'wichtigste'}]"
```

The result of the preceding HTTP call would be the following example sentences for the word "wichtigste," which all have a different meaning or context:

```
[
  {
    "normalizedSource": "key",
    "normalizedTarget": "wichtigste",
    "examples": [
      {
        "sourcePrefix": "The ",
        "sourceTerm": "key",
        "sourceSuffix": " problem at this stage is the ...",
```

```
      "targetPrefix": "Das ",
      "targetTerm": "wichtigste",
      "targetSuffix": " Problem zum jetzigen Zeitpunkt ist die ..."
    },
    {
      "sourcePrefix": "The ",
      "sourceTerm": "key",
      "sourceSuffix": " principle of the strategy is prevention is better
      ...",
      "targetPrefix": "Der ",
      "targetTerm": "wichtigste",
      "targetSuffix": " Grundsatz der Strategie lautet: \"Vorbeugen ist
      besser ..."
    },
    {
      "sourcePrefix": "The ",
      "sourceTerm": "key",
      "sourceSuffix": " action for 2007 is ...",
      "targetPrefix": "",
      "targetTerm": "Wichtigste",
      "targetSuffix": " Maßnahme im Jahr 2007 wird ..."
    },
    ...
  ]
 }
]
```

# Determine Sentence Length

In addition, the Translator API offers a feature for determining the length of a given
sentence as well as the ability to detect sentence boundaries. This can be extremely
helpful when you are confronted with a long piece of text, which you would like to
split into smaller pieces, before using the Text Analytics API's key phrase extraction
for obtaining certain key phrases within a long text. Given the following text "I was
really pleased with the nice service in the restaurant yesterday and I would definitely
come back. However, I did not like the fact, that the soup was almost cold whereas the

main dish was too hot to eat. But the ice cream dessert was perfect!" you could use the Translator API to detect the sentence boundaries programmatically using the following HTTP request:

```
curl -X POST "https://api.cognitive.microsofttranslator.com/
breaksentence?api-version=3.0" -H "Ocp-Apim-Subscription-Key:
yourTranslatorKey" -H "Content-Type: application/json" -d "[{'Text':'I was
really pleased with the nice service in the restaurant yesterday and I
would definitely come back. However, I did not like the fact, that the soup
was almost cold whereas the main dish was too hot to eat. But the ice cream
dessert was perfect!'}]"
```

The preceding command would result in the following response, whereas the sentence boundaries are determined as the character positions within the given input text:

```
[{"detectedLanguage":{"language":"en","score":1.0},"sentLen":[105,106,38]}]
```

These result values could now be used to split the text into smaller pieces and hand each of these pieces over to the Text Analytics API in order to extract the key phrases from each sentence individually, for example.

# Best Practices for Combining Cognitive Services Within a Chatbot

As you know have been introduced to the most used Azure Cognitive Services in conversational application scenarios individually, the following section will cover the best practices for combining some of these services. As said already, each of these Cognitive Services is basically offered as an API, which can be called from an application, no matter if it is a chatbot or any other solution; however, as the focus of this book is about conversational applications, the following scenarios will only cover conversational apps.

# LUIS and QnA Maker

When you build your very first chatbot, the best approach would be to start with a rather lightweight use case. Integrating QnA Maker into your chatbot could be one of these scenarios as you can quickly build up a chatbot based on a QnA Maker knowledge base. However, QnA is certainly not the only scenario which is used in a sophisticated chatbot solution. Therefore, LUIS is the perfect component when you need to cover QnA and task-based use cases in your chatbot. In such a scenario, it is important to design your natural language processing architecture accordingly. The best way to achieve this is to create a so-called Dispatch LUIS model, as outlined in Figure 3-59.

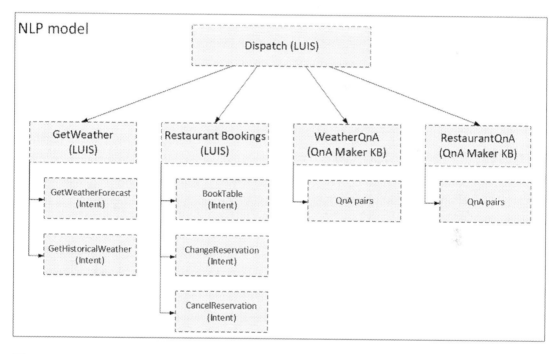

***Figure 3-59.*** *LUIS and QnA Maker reference architecture*

The concept of such an architecture is to create a top-level LUIS application which acts as a dispatch model for all underlying LUIS apps as well as QnA Maker knowledge bases. This dispatch application is used to determine which route is best suited when a new user query is received by your chatbot. The dispatch model basically acts as a router between all LUIS models and knowledge bases. The dispatch model basically contains an intent per LUIS application and knowledge base. When querying the dispatch model with a utterance, the dispatch app is then able to detect the matching intent which you then can use to forward your query either to the correct LUIS application or QnA Maker knowledge base, depending on if the utterance is something stored in one of your KBs or if it is an utterance which matches one your of LUIS apps' intents.

---

**Note**    To build up such a dispatch model, you need to use the Dispatch CLI tool which is not yet part of the Bot Framework CLI, but can be downloaded from `http://aka.ms/bot-dispatch`.

---

The above-outlined example in Figure 3-59 is something you may encounter when building a chatbot for a larger organization with multiple teams involved. In such a scenario, it may not always be the best approach to create one single LUIS application which covers all possible use cases. Therefore, you can split your language understanding model into more pieces, which offers the possibility for different teams to work on different LUIS apps independently, without affecting other parts of the language model.

## QnA Maker + Translator

Another best practice when building a multilingual chatbot especially is to make use of the Translator API, which is described earlier in this chapter, together with QnA Maker. This can be beneficial in cases where you build up knowledge bases of a larger scale for a chatbot which should target multinational scenarios. In such cases, you basically have three options:

1.  Build QnA Maker knowledge bases for each language which the chatbot should support.

2.  Build QnA Maker knowledge bases for the most important languages, and use the Translator API for all other languages.

3.  Build one QnA Maker knowledge base for the primary language of the chatbot, and use the Translator API for all other languages.

The first option obviously is the most time-consuming option, as it implies that you will need to build up a knowledge base per language you would like to support. Now if you develop a chatbot or conversational application, which is designed to support users from all around the world, you will need to build a lot of knowledge bases, which the chatbot then connects to. In many cases, the amount of effort for such an implementation is way too high, which is why this option is hardly ever chosen.

The effort for implementing the second option compared to the first one is a bit lower, as you do not need to build up a KB for each single language, but only for those languages which are high-priority ones. This option is suited best when you want to target specific countries especially but do not want to block certain languages. In such a situation, you could decide which languages should not be automatically translated but should be served from manually built up knowledge bases. User inquiries in these languages would get the answers from knowledge bases, which have been built up manually, targeting the specific language, which leads to better results in most cases.

## LUIS + Text Analytics

In many customer-facing chatbot examples, LUIS is used to detect the user's intent and start a specific routine or dialog within the conversation. Like in the example shown in Figure 3-59, there could be many routes within a conversation between the user and the chatbot, either getting information about the weather or managing a restaurant booking. In most scenarios the use of LUIS standalone might be sufficient as it detects the user's intent and acts accordingly. But especially in cases of an upset or angry user, it might be a good option to detect the user's sentiment and based on that decide if the chatbot should continue the conversation or escalate the conversation to a human.

Therefore, it might be a good option to integrate the Text Analytics API and LUIS together to cover these scenarios as well. As shown in Figure 3-60, if the conversation is following the happy path, the chatbot can continue the conversation and process the user's inquiry. Whereas in the critical path example as outlined, it is probably a better approach to hand off the conversation to a human agent, after the chatbot has detected a negative sentiment score in the user's message.

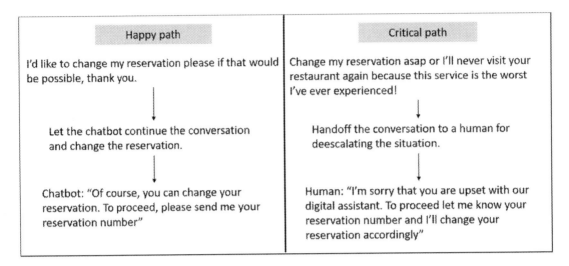

**Figure 3-60.** *LUIS and Text Analytics practical example*

# Summary

In this chapter you learned about the most important Azure Cognitive Services used in conversational applications such as chatbots. At first, we looked at the various services like LUIS, QnA Maker, Text Analytics, or Translator within the language category. These services are best suited in scenarios where users are having a conversation with your chatbot via written text. Additionally we learned how to create a new language understanding application using the web-based portal as well as the Bot Framework CLI as well as how to set up and populate a QnA Maker knowledge base via the QnA Maker portal and the CLI. The last part of this chapter covered some of the best practices for combining multiple Azure Cognitive Services when building a sophisticated chatbot solution, like multilingual scenarios or the ability to include multiple LUIS applications and QnA Maker KBs.

The next chapter will walk you through the core design principles and guidelines which should be taken into consideration when developing a chatbot. Moreover, the conversation flow concepts, user experience ideas, and visual elements will be covered for building exceptional chatbot solutions.

# CHAPTER 4

# Design Principles of a Chatbot

The overall design of your bot is the most important factor for your chatbot's success. Your aim as a developer is to build chatbots, which will be used by users. Therefore, you will need to think about how you could add benefits for users when using your chatbot instead of any other web application or other solutions. To make your users actually use your chatbot, the bot needs to solve the user's needs in the quickest and most importantly easiest way possible, compared to any other solution. An example would be the use case for booking a table in a restaurant. The "classic" approach would be as follows:

1. Open smartphone app or web browser, and navigate to a website for booking a restaurant table.

2. Enter the preferred city or restaurant.

3. Enter the restaurant style (e.g., Italian, Mexican, fast food, etc.).

4. Enter the number of guests.

5. Enter the time.

6. Pick from the list of available restaurants and confirm the booking.

Now in this case, your goal needs to be to ease this process, for example, by implementing a chatbot use case, which requires the user to fulfill the following steps:

1. Open the chatbot in a smartphone app or website.

2. Enter the phrase "I want to book a table for 2 on Sunday 6 p.m. at an Italian restaurant in Seattle."

3. Pick from the list of available restaurants and confirm the booking.

© Stephan Bisser 2021

S. Bisser, *Microsoft Conversational AI Platform for Developers*, https://doi.org/10.1007/978-1-4842-6837-7_4

In the preceding example, you see that the booking process done with a chatbot is far easier and less time-consuming compared to the usual approach within an app or a website. Therefore, users will likely use your chatbot over an app or a website, as they can save time and still will be getting the expected result. In order to build a successful chatbot, there are many things which you need to take into consideration; therefore, this chapter will cover all important aspects for designing a successful chatbot.

# Personality and Branding

First of all, the key to your bot's success is its personality. If your bot has a unique personality, users will likely want to have a conversation with your bot as it makes the overall chat experience more human and natural. Therefore, you should build a brand for your chatbot. Usually your brand should include the following key attributes to stand out from others:

- **Name**: Do not call your bot "Chatbot XYZ," but give it an appropriate name so that users can relate to your bot.

- **Slogan**: Give your bot a short slogan or introduction phrase to make people aware of what your bot is capable of.

- **Picture**: Show people how your bot looks like, may it be a robot-like image or an illustration of a human to let people see who they are having a conversation with.

Your chatbot is basically the representation of yourself or your company to your users, so you need to make sure that the personality of your bot is reflecting your core values. Imagine you are building a chatbot for a banking institution where all employees who are serving customers at the counters are wearing suits and ties. In this situation the chatbot should be aligned to the bank's mindset to be polite and serve customers in a respectful way. Therefore, it would make no sense to build a bot speaking in a witty style targeting the younger generation as this is not the primary target group of a bank in many cases. It would make more sense to align the personality and the language of your chatbot to the overall representation of the company which is utilizing the bot. In contrast to this, if you are building a chatbot for a young startup company, focusing on selling lifestyle goods, a witty personality would be totally appropriate as this may also be how the company wants to be represented to its users or customers.

Before developing a chatbot, you should focus on building or at least outlining your bot's personality to fit the desired use cases and, more importantly, its audience. This is in many cases not a task which has to be done by a developer alone but requires business users who know the target audience and can therefore draw a clear picture of the personality of your chatbot. Needless to say, this process involves persona definitions and key user analysis in most cases, which is why this is certainly not a task which should be done by developers only. You always need to keep in mind that your chatbot is another virtual representation of your company or your services. Therefore, it should be tightly aligned to the overall appearance of your company or offerings.

In general people prefer a chatbot to respond rather humanlike than robotic. Therefore, it is essential to keep your target audience in mind. If the target audience is generation Z for instance, the chatbot could of course respond with emojis and things that this generation is used to when chatting with their peers. In contrast to that, when your target audience is the baby boomer generation, your chatbot should talk the language these people are used to. One appropriate method to investigate on this is the use of so-called personality cards, which basically describe the chatbot's personality in different situations.

---

**Note**    The primary goal here is that the chatbot personality is consistent through the entire conversation to be adapted by the end users.

---

# Greeting and Introduction

One important part of your chatbot is how it introduces itself to your users. This should of course be aligned to the overall personality or branding of your chatbot, to offer a consistent chatbot experience. If you, for example, build a chatbot with a friendly personality, try to express this friendliness right from the beginning of the conversation. On the other hand, if your bot should be rather formal, the greeting or introductory phrase should be targeted toward that.

Furthermore, you need to keep in mind that there is only one chance to make a positive impression. If your users would like the introduction and greeting part of the chatbot, it is more likely that they will continue the conversation with it. Thus, it is more likely that they will consume your services offered through the bot or visit your website to gain more information about your offerings. So as the first impression is crucial here, you need to make sure that the greeting leaves a positive impression on users. If that is the case, the chances are much higher that your chatbot will actually be used. Moreover, it is much more likely that you can utilize your bot to promote your offerings through your chatbot.

To give you an example of what should be avoided in the greeting part of the conversation, Figure 4-1 demonstrates how a rather bad greeting situation could look like. First of all, the most important part here is that the bot does not greet the user first. This comes in somewhat unfriendly, which is why you should always try to implement what's called "proactive greeting." This is a concept of greeting the user first, before the user had the chance to even type in "hi." Second, the chatbot asks a quite open question "How can I help you?" Typically, these types of questions lead to user questions which are not precise enough, as the user does not know immediately what to ask. Thus, this bad greeting example should be avoided, as it does not introduce the main abilities of the chatbot immediately. Therefore, users might not know what to ask and will close the chat soon.

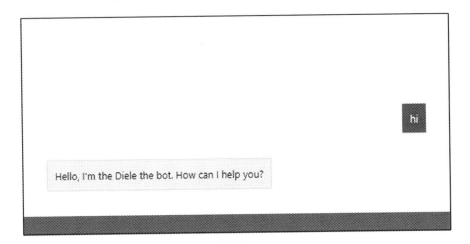

***Figure 4-1.*** *Bad chatbot greeting example*

In contrast to the bad example, Figure 4-2 shows a better example of how to handle greetings. In this sample, you see that the bot greets the user first, which is somewhat polite. Moreover, the bot does not only greet the user, but it introduces itself, stating the main abilities and use cases covered. This is of advantage, as the user gets directly the explanation what he or she can ask the bot or what the bot is capable of doing in general. Additionally, the greeting part includes not only text but also some buttons, which the user can press directly. These buttons act as some kind of navigation helper within the conversation, to guide the user through the conversation. This has two major advantages; first of all, the user will be able to move along within the conversation faster as there is no need of typing every phrase manually into the chat. Second, the bot will suggest possible paths within the dialog or conversation tree, which limits down the possibilities the user has in terms of asking questions. This, in turn, minimizes problems with language understanding components, as the message is already predefined by the button.

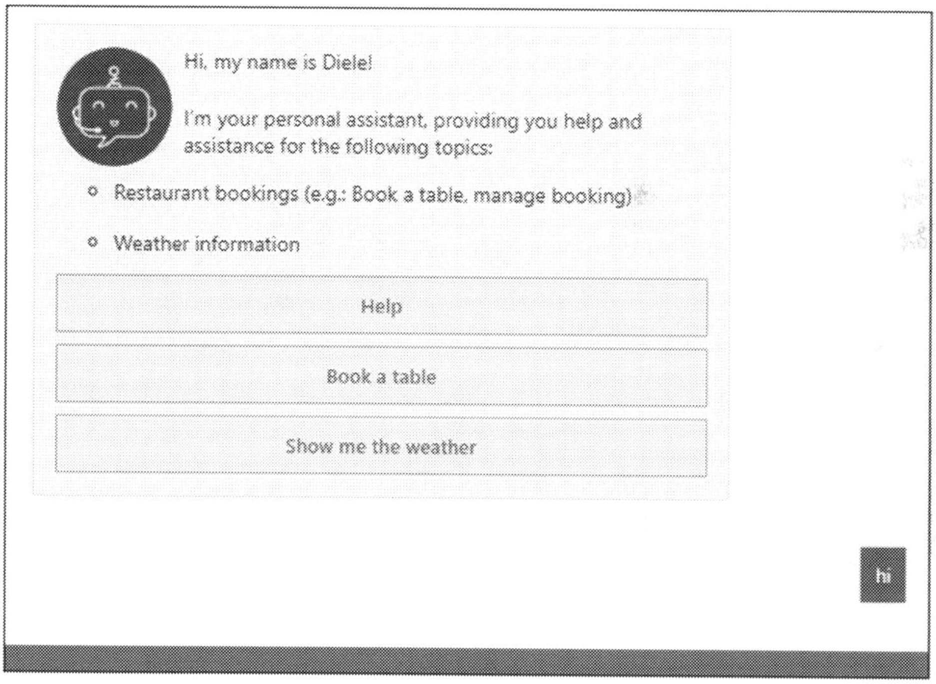

*Figure 4-2.*  *Good chatbot greeting example*

You should also include proactive greeting mechanisms. This will help your users get to know the chatbot better. As Figure 4-2 illustrates, the bot introduces itself to the user with a greeting card. This card basically consists of the bot's name, the icon, and most importantly the motivation of the chatbot. This way your bot can tell the users what its strengths are, before the user needs to ask the bot anything. As with this example, the bot should also give the user a chance to start the conversation quickly, most commonly done with buttons. This way, the user does not even need to type something into the chat, but instead can click the button to start a specific dialog within the conversation.

The greeting is a crucial part within the conversation life cycle, as it is decisive of the reaction of the users. Remember that the first impression always counts, so if the bot leaves a good first impression, users are more willing to continue the conversation with the bot. This in turn will increase the usage numbers of your bot and may result in more satisfied customers in the end. Therefore, the personality of your bot is again important, which should be expressed in the greeting message also. If your users receive a warm and appropriate greeting by the chatbot, they will use the chatbot.

# Navigation (Menu)

Another important part of a chatbot conversation is the conversational navigation. In contrast to an app or a website, a chatbot does not have a hamburger menu or a navigation bar which offers the possibility to jump to specific points within the conversation. Therefore, you need to think about how to guide users through the conversation. Additionally, the management of misunderstandings or dead ends should be covered as well. Therefore, it is a good approach to integrate a navigation strategy, which the bot communicates toward the users. For example, Figure 4-3 demonstrates a good approach of call attention to how users may receive help within a conversation, by simply responding with "Help" in the chat. This way, users will see that in the very first message within the conversation and know, therefore, whenever something is unclear or if they do not know how to proceed, they can simply type in help to receive support to continue the conversation.

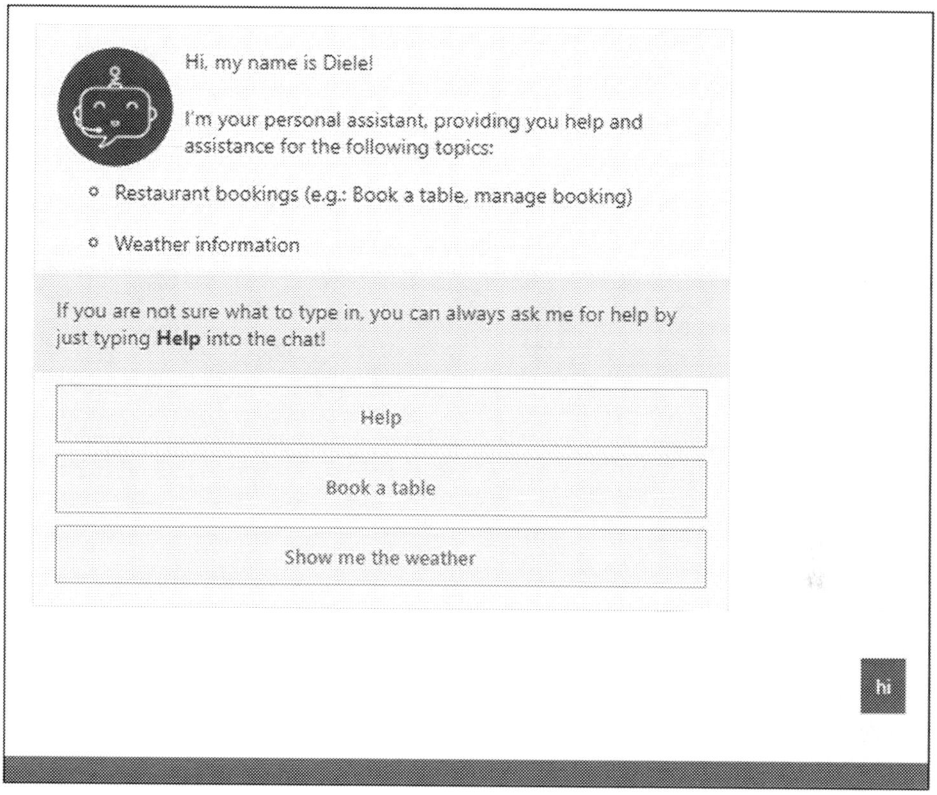

*Figure 4-3.* *Greeting card navigation guidance sample*

But this option is not the only option within a conversation, as users might not always remember the description within the greeting message; therefore, you should also aim to offer help whenever you think that a user is not certain on how to proceed the conversation. This can basically be done by using natural language processing to detect whenever a user is uncertain about something. In such situations the chatbot should provide the user with some sort of assistance. To give you an example of how this should not be handled, Figure 4-4 outlines a conversation sequence which should not occur in practice. In this example, the user might think that the chatbot is not intended to fulfill the user's needs and might therefore end the conversation immediately.

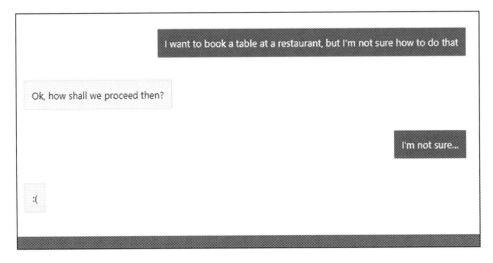

***Figure 4-4.*** *Bad navigation/help example*

To overcome such an issue, Figure 4-5 outlines an example of a good approach for dealing with an uncertain user. In this example, the chatbot identifies that the user does not exactly know how to continue the conversation, which triggers an intent within the language understanding component within the chatbot, leading to the question if the user wants to book a table or needs general help. As the user wants to get general information about the chatbot's tasks, a card gets sent out to the user, describing what exactly the chatbot is capable of doing for the user along with visual elements, in the form of buttons, to guide the user to the right part within the dialog.

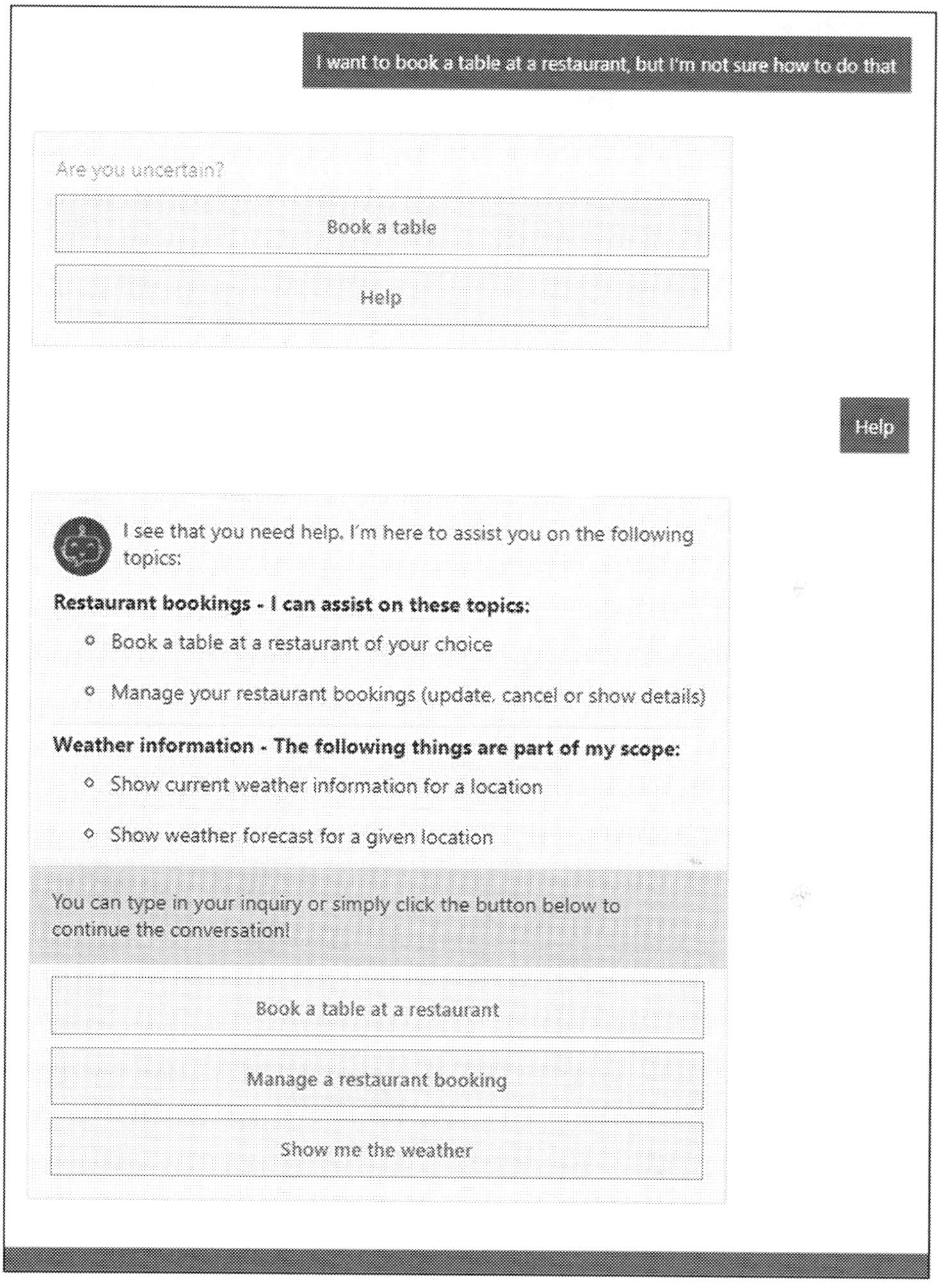

I want to book a table at a restaurant, but I'm not sure how to do that

Are you uncertain?

Book a table

Help

Help

I see that you need help. I'm here to assist you on the following topics:

**Restaurant bookings - I can assist on these topics:**

○ Book a table at a restaurant of your choice

○ Manage your restaurant bookings (update, cancel or show details)

**Weather information - The following things are part of my scope:**

○ Show current weather information for a location

○ Show weather forecast for a given location

You can type in your inquiry or simply click the button below to continue the conversation!

Book a table at a restaurant

Manage a restaurant booking

Show me the weather

***Figure 4-5.*** *Good navigation/help example*

As the preceding example shows, the chatbot offers assistance in uncertain situations in the form of navigation buttons, which the user can click to trigger a specific dialog within the conversation tree. Therefore, the conversation can continue fluently, and the user gets guided to the right point within the conversation.

# Conversation Flow

A conversation usually consists of multiple dialogs which, together, form a dialog tree. Such a dialog tree is basically a representation of your chatbot's use case. In practice, it is a good approach to model the dialog tree before actually starting developing or building your bot. By doing this, you basically sketch out the bot's behavior and routines on paper in the form of a flow chart or similar options, which acts as your conversation blueprint for the later development stage. As mentioned in Chapter 2, dialogs are the core of your chatbot, as they are the direct conversational mechanism users will be offered when having a conversation with your bot. While designing your conversation flow, you may also consider how to handle interruptions properly. Because it could happen that the user might want to switch the context in the middle of a dialog, therefore, you may also think about how to manage these kinds of situations, as well as allowing the conversation to resume at the point at which the interruption occurred.

As shown in Figure 4-6, the dialog tree of your chatbot may consist of many different dialogs. Each dialog is responsible for handling a different use case or part of the conversation. The parent dialog in the Microsoft Bot Framework is called the *RootDialog* or *MainDialog*. This is basically the starting point of every conversation between a user and your chatbot. The RootDialog usually includes a LUIS recognizer, which is responsible for determining the user's intent to trigger a specific child dialog.

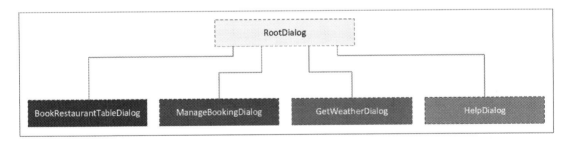

*Figure 4-6.*  *Dialog tree overview*

Before starting to develop your chatbot, it is usually a good idea to model the dialogs as seen in Figure 4-6 to get an idea which dialogs and conversational mechanisms are needed when building your bot. To assist developers and chatbot designers in this process, the Bot Framework CLI offers a tool for turning chat files, which are markdown representatives into transcript files to be used within the Bot Framework Emulator for representation purpose. A simple .chat file may look as follows:

```
user=stephan
bot=diele

user: hi
bot: Hello, I'm the Diele the bot. How can I help you?
```

In this example, the preceding markdown content is stored in a file called *04_greeting_bad.chat*. After the file has been created, you can run the following Bot Framework CLI command within your terminal to convert this .chat file into a .transcript file:

```
bf chatdown:convert --in .\04_greeting_bad.chat -o .\
```

The transcript file will include many additional parameters than the chat file, like the conversation and conversation member IDs, the activity types, and the messages. Now this generated transcript file can be opened from the Bot Framework Emulator, which will basically show you how this very simple conversation would look like when displayed in a real-world scenario, as shown in Figure 4-7.

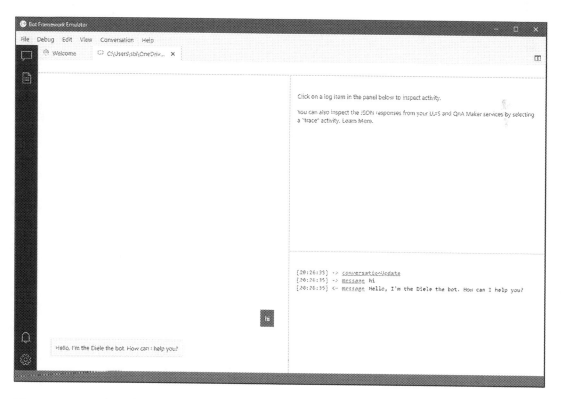

***Figure 4-7.*** *Chatdown example 01*

But you can not only use plain text in .chat files but also cards, which will be discussed in depth in a later section in this chapter. The use of cards in chat files is shown in the example here:

```
user=stephan
bot=diele

bot: [Attachment=cards\04_greetingCard.json adaptivecard]
user: hi
```

After executing the bf chatdown:convert command again and passing in the preceding sample chat file, the conversation rendered in the Bot Framework Emulator will look as demonstrated in Figure 4-8.

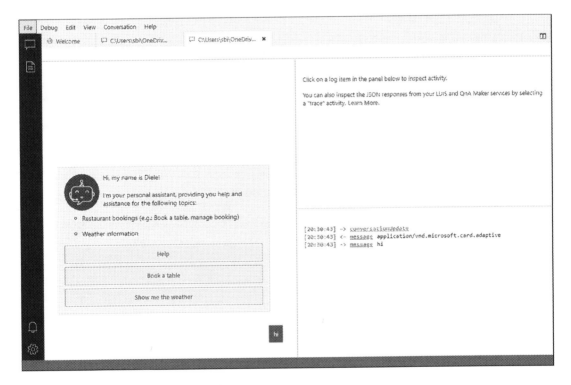

**Figure 4-8.** *Chatdown example 02*

But Chatdown is not only intended to be used as shown above the sketch out and review just one turn within a conversation. You can also use Chatdown to design complete dialogs and conversation paths within your chatbot. For example, you could build up a .chat file for the *BookRestaurantTableDialog* as outlined in Figure 4-6.

This dialog aims at booking a table for a user through the chatbot. Therefore, the chatbot must know the preferred location, the date and time, as well as the number of guests at least to be able to book a restaurant table. To express this using a chat file which can then be converted into a transcript file using the Bot Framework CLI again, you could use the following example:

```
user=stephan
bot=diele

user: book a table
bot: Did you have a location in mind?
bot: [Attachment=cards/04_askForLocationCard.json adaptivecard]
user: How about seattle?
bot: Did you have a date in mind?
user: tomorrow
bot: What time?
user: 3PM
bot: How many guests?
user: 3
bot: Ok. Should I go ahead and book a table for 3 at seattle for tomorrow
at 3PM?
bot: [Attachment=cards/04_tableConfirmationCard.json adaptivecard]
user: yes
bot: [Typing][Delay=3000]
Your table is booked. Reference number: #K89HG38SZ
```

After converting the preceding chat file into a transcript file, the output within the Bot Framework Emulator is shown in Figure 4-9. As you can see in this example, it is possible to design complete dialogs or even a whole conversation path using only a few lines of markdown. Thus, even nondevelopers could use Chatdown and the .chat file notations to design chatbot conversations for reviews and demonstration purposes, before handing off to developers for actually developing the chatbot's logic and dialogs.

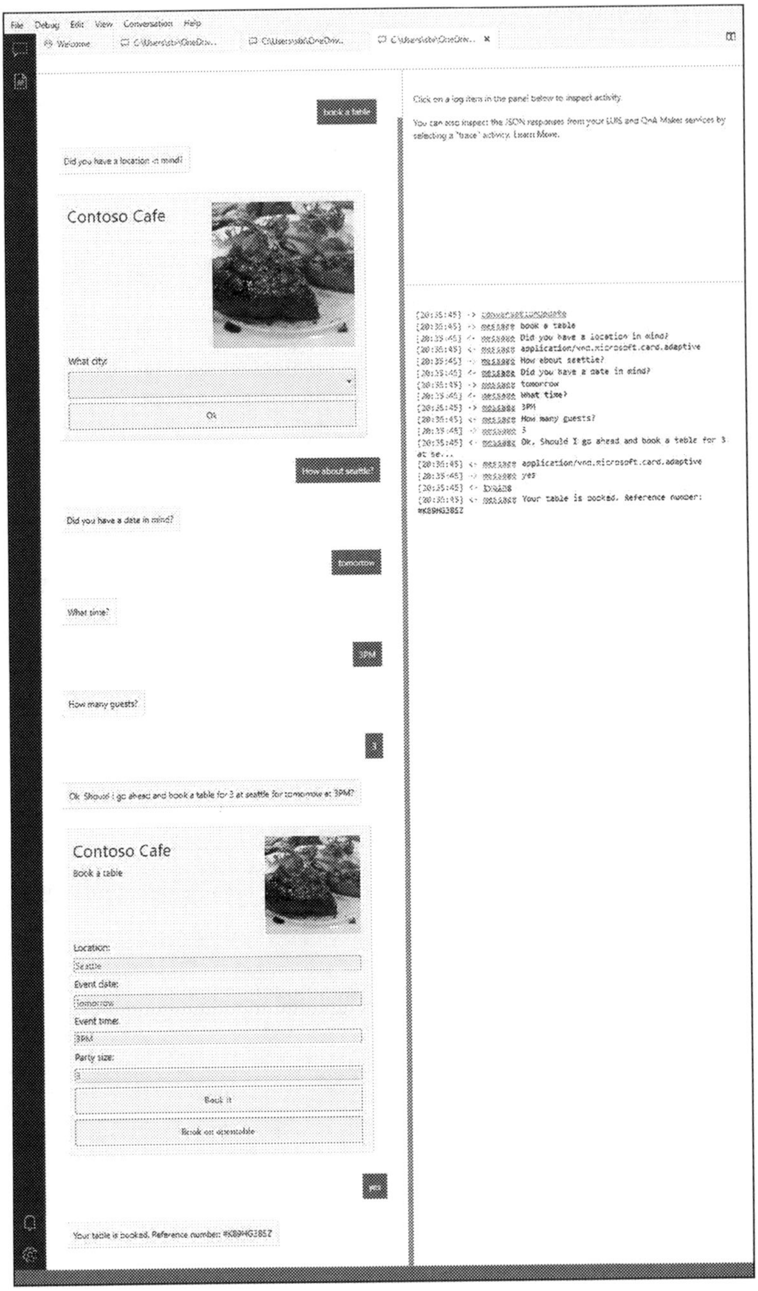

***Figure 4-9.*** *Chatdown example 03*

# Adaptive Dialogs

Within the Bot Framework, there are nowadays two concepts for implementing dialogs, the dialogs library, which is the concept used for many Bot Framework SDK versions now, and the newly added adaptive dialogs concept. The main difference between them is that adaptive dialogs offer more flexibility in terms of handling conversational events, such as interruptions or context switches. The main idea behind adaptive dialogs is to offer a more event-based approach for building dialogs, which contain both the conversation parts, like language understanding and generation, and the actions within the dialog. As this new declarative model is used to provide developers with a set of patterns for building chatbots, not necessarily by developing code, the focus of this book is mainly around describing this adaptive dialog approach.

---

**Note**   If you want to learn more about the dialogs library within the Bot Framework SDK, visit `https://docs.microsoft.com/en-us/azure/bot-service/bot-builder-concept-dialog`.

---

Adaptive dialogs contain a set of event handlers which are called triggers, as shown in Figure 4-10. A trigger itself can contain a condition, to determine when the trigger gets triggered. Within each trigger, there are one or more actions which will be executed upon running the trigger. The event handler basically handles all events happening within the conversation and evaluates which trigger should be run.

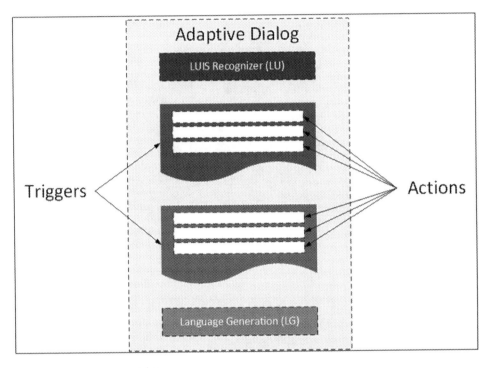

**Figure 4-10.**  *Adaptive dialogs concept*

## Adaptive Dialog Triggers

The Bot Framework includes many different trigger types which are used for specific scenarios, as outlined in Table 4-1, ranging from recognizer events to dialog and activity events, up to message events.

***Table 4-1.***  *Adaptive dialogs – trigger types (Microsoft, 2020)*

| Trigger Type | Description |
| --- | --- |
| **Base trigger** | The *OnCondition* trigger is the base trigger that all triggers derive from. When defining triggers in an adaptive dialog, they are defined as a list of *OnCondition* objects. |
| **Recognizer event triggers** | Recognizers extract meaningful pieces of information from a user's input in the form of intents and entities, and when they do, they emit events. For example, the *recognizedIntent* event fires when the recognizer picks up an intent (or extracts entities) from a given user utterance. You handle this event using the *OnIntent* trigger. The following triggers are available in this type:<br>• *OnChooseIntent*<br>• *OnIntent*<br>• *OnUnknownIntent*<br>• *OnQnAMatch* |
| **Dialog event triggers** | Dialog triggers handle dialog-specific events that are related to the "life cycle" of the dialog. There are currently six dialog triggers in the Bot Framework SDK, and they all derive from the *OnDialogEvent* class. The following triggers are available in this type:<br>• *BeginDialog*<br>• *CancelDialog*<br>• *EndOfActions*<br>• *Error* |
| **Activity event triggers** | Activity triggers let you associate actions to any incoming activity from the client such as when a new user joins and the bot begins a new conversation. Additional information on activities can be found in Bot Framework Activity schema.<br>All activity events have a base event of *ActivityReceived* and are further refined by their activity type. The Base class that all activity triggers derive from is *OnActivity*. The following triggers are available in this type:<br>• *ConversationUpdate*<br>• *EndOfConversation*<br>• *Event*<br>• *Handoff*<br>• *Invoke*<br>• *Typing* |

*(continued)*

***Table 4-1.*** (*continued*)

| Trigger Type | Description |
| --- | --- |
| **Message event triggers** | Message event triggers allow you to react to any message event such as when a message is updated (*MessageUpdate*) or deleted (*MessageDeletion*) or when someone reacts (*MessageReaction*) to a message (e.g., some of the common message reactions include like, heart, laugh, surprised, sad, and angry reactions). The following triggers are available in this type. |
| | Message events are a type of activity event, and, as such, all message events have a base event of *ActivityReceived* and are further refined by activity type. The Base class that all message triggers derive from is *OnActivity*. The following triggers are available in this type:<br><br>• *Message*<br>• *MessageDeletion*<br>• *MessageReaction*<br>• *MessageUpdate* |
| **Custom event triggers** | You can emit your own events by adding the *EmitEvent* action to any trigger; then you can handle that custom event in any trigger in any dialog in your bot by defining a custom event trigger. A custom event trigger is the *OnDialogEvent* trigger that in effect becomes a custom trigger when you set the Event property to the same value as the *EmitEvent's EventName* property. The following triggers are available in this type:<br><br>• *OnDialogEvent* |

The recognizer event triggers include recognizers which are used to determine the user's intents and entities. The *OnChooseIntent* trigger within the recognizer event triggers is fired, when there are multiple intents recognized by multiple recognizers within the bot. Therefore, this trigger basically handles the logic on how to choose the correct intent to continue within the dialog stack. The *OnIntent* trigger within this trigger type gets executed when a specific intent has been recognized to allow the defined actions to be run. In contrast to that, the *OnUnknownIntent* trigger handles situation when there is no intent recognized by any of the *OnIntent* triggers. And the last trigger within the recognizer event triggers is the *OnQnAMatch* trigger. This trigger is run whenever there has been a *QnAMatch* intent received by the *QnAMakerRecognizer*, which means that a match has been identified between the user's input and a QnA pair within a QnA Maker knowledge base.

The dialog event triggers type consists of triggers which are used to handle specific events during the life cycle of a dialog. The *OnBeginDialog* trigger includes actions which are taken when a specific dialog starts. The *OnCancelDialog* trigger is used to avoid a dialog from being canceled. The *OnEndOfActions* trigger defines which actions should be executed after all actions of a dialog have been run. And the *OnError* trigger handles specific error situations within a dialog.

Within the activity event triggers types, the *ConversationUpdate* trigger is run when a user starts a conversation with a bot. The *EndOfConversation* trigger gets executed when the conversation between the user and the bot ends. The *Handoff* trigger is used to perform a human handoff action to loop a human into the conversation with the user. The *Typing* trigger is used to determine if the user is currently typing.

The *MessageReceived* trigger within the message event triggers type is used to execute actions when a new message from the user has been received by the bot. In contrast to this, if a message from the user has been deleted, the *MessageDelete* trigger gets executed. When a user reacts to a specific message with a like, for example, the *MessageReaction* trigger is run to handle such situations and perform necessary actions based on that. The *MessageUpdate* trigger can be used to handle an update of a user message.

## Adaptive Dialogs Actions

Besides triggers, adaptive dialogs also contain *actions*. These actions are basically the steps or turns which are executed when a dialog is triggered. Actions are mainly focusing on preserving the conversation flow, like sending messages or querying a knowledge base to answer a user's question or execute another kind of task, like validating a user's input to continue the conversation. As there are many different actions available in adaptive dialogs, Table 4-2 briefly describes the most commonly used actions.

**Note**    To get a list of all supported actions in adaptive dialogs, please conduct
`https://docs.microsoft.com/en-us/azure/bot-service/adaptive-dialog/adaptive-dialog-prebuilt-actions`.

***Table 4-2.*** *Adaptive dialogs action types (Microsoft, 2020)*

| Action Type | Action Name | Action Description |
|---|---|---|
| **Activities** | SendActivity | Sends an activity, such as a response to a user. |
| **Activities** | Get activity members | Gets a list of activity members and saves it to a property in memory. |
| **Conditional statements** | IfCondition | Runs a set of actions based on a Boolean expression. |
| **Conditional statements** | SwitchCondition | Runs a set of actions based on a pattern match. |
| **Conditional statements** | ForEach | Loops through a set of values stored in an array. |
| **Dialog management** | BeginDialog | Begins executing another dialog. When that dialog finishes, the execution of the current trigger will resume. |
| **Dialog management** | EndDialog | Ends the active dialog. Use when you want the dialog to complete and return results before ending. Emits the EndDialog event. |
| **Dialog management** | RepeatDialog | Used to restart the parent dialog. |
| **Manage properties** | EditArray | Performs an operation on an array. |
| **Manage properties** | SetProperties | Sets the value of one or multiple properties at once. |
| **Access external resources** | BeginSkill | Begins a skill and forward activities to the skill until the skill ends. |
| **Access external resources** | HttpRequest | Makes an HTTP request to an endpoint. |
| **Access external resources** | CodeAction | Calls custom code. The custom code must be asynchronous, take a dialog context and an object as parameters, and return a dialog turn result. |

# Adaptive Dialogs Memory Scopes

All the abovementioned actions can be used to execute different tasks. In addition to that, the actions within dialogs also have access to the bot's memory layer. This enables developers to manage the bot's state using adaptive dialogs similar to the concept within the dialogs library in the Bot Framework SDK. The memory within adaptive dialogs is split into different *memory scopes*. Each scope is offered to serve a different purpose, like persisting user or conversation data or accessing values within a specific dialog. All memory scopes within adaptive dialogs are described in Table 4-3.

*Table 4-3.* *Adaptive dialogs memory scopes (Microsoft, 2020)*

| Scope Name | Scope Description |
| --- | --- |
| User scope | User scope is persistent data scoped to the ID of the user you are conversing with. Examples would be<br><br>• user.name<br>• user.address.city |
| Conversation scope | Conversation scope is persistent data scoped to the ID of the conversation you are having. Examples would be<br><br>• conversation.hasAccepted<br>• conversation.dateStarted |
| Dialog scope | Dialog scope persists data for the life of the associated dialog, providing memory space for each dialog to have internal persistent bookkeeping. Dialog scope is cleared when the associated dialog ends. Examples would be<br><br>• dialog.orderStarted<br>• dialog.shoppingCart |
| Turn scope | The turn scope contains nonpersistent data that is only scoped for the current turn. The turn scope provides a place to share data for the lifetime of the current turn. Examples would be<br><br>• turn.bookingConfirmation<br>• turn.activityProcessed |
| Settings scope | This represents any settings that are made available to the bot via the platform-specific settings configuration system, for example, if you are developing your bot using C#, these settings will appear in the appsettings.json file. Examples would be<br><br>• settings.QnAMaker.hostname<br>• settings.QnAMaker.endpointKey<br>• settings.QnAMaker.knowledgebaseId |

*(continued)*

***Table 4-3.*** (*continued*)

| Scope Name | Scope Description |
|---|---|
| **This scope** | The this scope pertains the active action's property bag. This is helpful for input actions since their life type typically lasts beyond a single turn of the conversation. Examples would be<br><br>• this.value (holds the current recognized value for the input)<br>• this.turnCount (holds the number of times the missing information has been prompted for this input) |
| **Class scope** | This holds the instance properties of the active dialog. You reference this scope as follows: ${class.<propertyName>}. |

Putting this all together, an example of the dialog tree outlined in Figure 4-6 designed with the adaptive dialogs concept can be seen in Figure 4-11. This example shows the implementation of the *RootDialog* along with its different triggers of type *OnIntent*. Each of these triggers includes a *BeginDialog* action to execute the appropriate child dialog based on the user's intent. Within the child dialogs, there may again be certain triggers along with corresponding actions to fulfill the dialog accordingly. All dialogs and actions share the same memory layer, which means each dialog can access the described memory scopes accordingly, to enable appropriate state management. This example dialog tree will also be used in the next chapters to actually develop a chatbot designed for handling these conversational tasks.

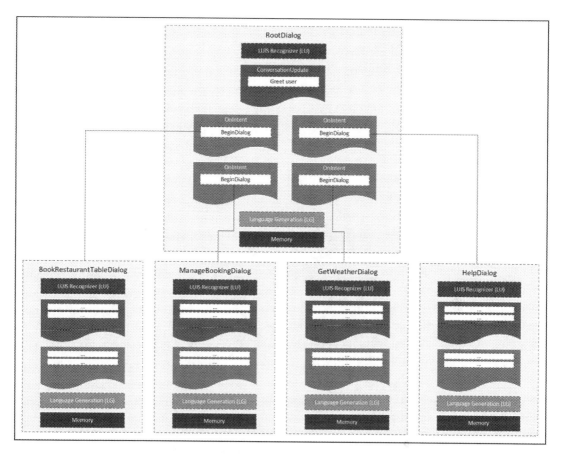

***Figure 4-11.*** *Adaptive dialog tree example*

# User Experience

The user experience for a chatbot is somehow different to the "classic" user experience of an app or a website as you are building a conversation user interface (short, CUI) and not a graphical user interface (short, GUI). As people have different kinds of conversational habits, the task of designing the user experience for a chatbot could be sometimes challenging. Therefore, it is vital, as mentioned earlier in this chapter, to know the target audience of your chatbot. Younger people are used to utilize conversational platforms like Facebook, WhatsApp, and Slack. Thus, these people are used to communicate with others via instant messaging and chat, whereas older generations may not be used to this kind of communication platforms as they grew up in different eras where they communicated with each other differently. In fact, younger people are familiar with emojis and rich media like images and videos within conversation, which may not be the case for older generations. Therefore, before developing your chatbot, you should have a clear understanding of the target group.

In general, however, it is a good practice to develop a chatbot which is not only communicating with its users via text-only messages. Of course, this requires that the channel where your chatbot will be accessible through supports other types of messages and rich media attachments. For instance, a bot used within text messages (SMS) will only be able to send and receive text messages and not images and videos. Yet, most communication channels, which are supported by the Azure Bot Service, support rich media attachments. Hence, your chatbot should be designed to include these conversational mechanisms, to positively influence the conversation flow.

Additionally, a good chatbot user experience includes the concept of state management. The state is basically the bot's brain and is therefore used to let the bot recall the conversation history with a user. This is extremely important to design a good conversation flow as the bot can handle interruptions much better when it knows where the interruption has happened and how to resume the conversation to the point at which an interruption or context switch has taken place. Moreover, by including proper state management, you have the chance to design a more human conversation behavior as the bot will be able to remember certain things from previous conversations with a user, which enhances the conversation flow as well. In our example, the chatbot is serving restaurant booking requests which also includes the manage bookings. Therefore, it would be advantageous if the user would not need to provide the complete information regarding a restaurant booking to update or cancel the booking. Instead, the bot should be able to remember the user's bookings and just ask for the booking reference for instance to continue the user's request. Thus, state management introduces the bot's capability of remembering information from previous conversations.

# Rich Media Attachments

Apart from the bot's capability of remembering parts of previous conversations, another important factor for a good user experience is to include rich media attachments apart from text. As stated earlier, it is dependent of which channel your bot is deployed to, but in general it is a good approach to include conversation parts like images, links, videos, or even emojis within the conversation with the user. Therefore, the Microsoft Bot Framework supports many different options of adding rich media attachments to a bot's message. In C# bots you may use the following code snippet to easily send a message which contains an image to the user by adding an inline attachment to a message:

```
msg = MessageFactory.Text("Here you will see the dialog tree!");
msg.Attachments = new List<Attachment>() { GetImageAttachment() };
private static Attachment GetImageAttachment()
{
    var imgPath = Path.Combine(Environment.CurrentDirectory, @"img",
    "dialogTree.png");
    var img = Convert.ToBase64String(File.ReadAllBytes(imgPath));
    return new Attachment
    {
        Name = @"img\dialogTree.png",
        ContentType = "image/png",
        ContentUrl = $"data:image/png;base64,{img}",
    };
}
```

The important thing to note here is that rich media attachments should enhance the conversation between the bot and the user. Responding with messages which include images and other media is beneficial when you want to demonstrate certain things like a menu card. Of course, you could send a text-based message to the user containing the information about the menu, but in most scenarios an image of a menu card is somewhat more professional and most importantly user-friendly.

## Cards as Visual Elements

In addition to using inline attachments, it is generally a good approach to use what is called a *card* as an introductory message to start the conversation with a user proactively for instance, as demonstrated earlier in this chapter. Cards within the Bot Framework can include different types of rich user controls as well as visual or audio items within one single message. As Table 4-4 describes, there are many different card types supported within the Microsoft Bot Framework, which can be used for different scenarios, like displaying an animation, a video, or image, playing an audio file, or logging in the user. As this chapter will not cover all supported card types in detail, we will focus on the most commonly used card types in the next few paragraphs.

***Table 4-4.*** *Support card types in the Microsoft Bot Framework (Microsoft, 2018)*

| Card Type | Card Description |
| --- | --- |
| **Adaptive Card** | An open card exchange format rendered as a JSON object. Typically it is used for cross-channel deployment of cards. Cards adapt to the look and feel of each host channel. |
| **Animation card** | A card that can play animated GIFs or short videos. |
| **Audio card** | A card that can play an audio file. |
| **Hero card** | A card that contains a single large image, one or more buttons, and text. Typically it is used to visually highlight a potential user selection. |
| **Thumbnail card** | A card that contains a single thumbnail image, one or more buttons, and text. Typically it is used to visually highlight the buttons for a potential user selection. |
| **Receipt card** | A card that enables a bot to provide a receipt to the user. It typically contains the list of items to include on the receipt, tax and total information, and other text. |
| **Sign-in card** | A card that enables a bot to request that a user sign in. It typically contains text and one or more buttons that the user can click to initiate the sign-in process. |
| **Suggested action** | Presents your user with a set of CardActions representing a user choice. This card disappears once any of the suggested actions is selected. |
| **Video card** | A card that can play videos. Typically it is used to open a URL and stream an available video. |
| **Card carousel** | A horizontally scrollable collection of cards that allows your user to easily view a series of possible user choices. |

# Hero Cards

In contrast to attaching an image or video to a message, hero cards offer the possibility to combine text, images, and videos as well as buttons within one single message. This is often used to create greeting or help cards or any other kind of message where you want to include text and media in one message. Hero cards also give users the possibility to continue the conversation with a click of a button, instead of typing in text into the chat. This is especially helpful on mobile phones, as it makes the conversation more fluent and faster. A hero card basically consists of the parts which are outlined in Figure 4-12.

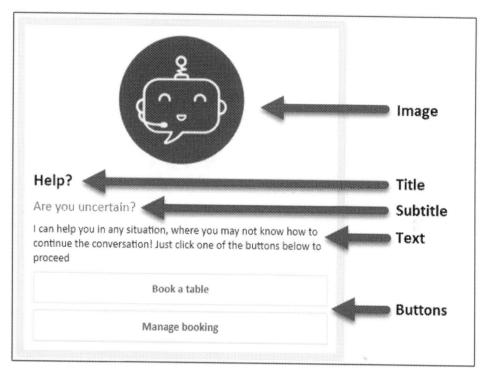

*Figure 4-12.   Hero card example*

Within C# bots, you could use the following lines of code to generate a hero card and send it out to users within a conversation:

```
var attachments = new List<Attachment>();
var reply = MessageFactory.Attachment(attachments);
reply.Attachments.Add(GetHeroCard().ToAttachment());
await stepContext.Context.SendActivityAsync(reply, cancellationToken);
public static HeroCard GetHeroCard()
{
    var heroCard = new HeroCard
    {
        Title = "Help?",
        Subtitle = "Are you uncertain?",
```

```
Text = " I can help you in any situation, where you may not know
how to continue the conversation! Just click one of the buttons
below to proceed",
Images = new List<CardImage> { new CardImage("https://bisser.io/
images/bisser_io_red.png") },
Buttons = new List<CardAction>
{
    new CardAction(ActionTypes.ImBack, title: "Book a table",
    value: "Book a table"),
    new CardAction(ActionTypes.ImBack, title: "Manage booking",
    value: "Manage booking"),
},
return heroCard;
}
}
```

## Card Carousels

The preceding example showed you how to create and send out one instance of a hero or Adaptive Card. But in many situations, you want to visualize multiple cards within one single message. Therefore, you can use the *AttachmentLayout* property within the hero card class to create a carousel of cards as shown in Figure 4-13. This way, you can add multiple cards to the same activity. The user who receives this carousel of cards can then swipe to the right or left to view the cards which are attached to the message.

**Figure 4-13.** *Hero card carousel example*

This functionality is very helpful in scenarios where you want to show the same kind of visualization type with different data in one message, for example, as shown in the preceding image, showing a user's restaurant bookings. Of course, you could go ahead and send out separate messages and cards for each booking, but this would maybe lead to the issue of sending out a great number of separate messages after each other. The user then needs to scroll and browse through those messages within the chat window, which can, especially on mobile phones, somehow be difficult which could annoy users. Therefore, using the option of sending out carousel cards enhances the user experience as it is easier to view and navigate between cards, which is why it should be used in appropriate situations.

## Suggested Actions

In contrast to buttons within cards, which the user can tap to continue the conversation, suggested actions are buttons which are displayed close to the chat input text box, which will disappear after the user has selected one of these buttons or entered text. These suggested actions are a good use to offer users of answering questions by tapping a button instead of entering text. The main advantage of using suggested actions over buttons in cards is in fact that the buttons in cards remain visible in the chat window. Therefore, it could happen that a user clicks a button from a previous message again, which might cause an interruption or context switch as the bot is already in a different part within the dialog tree. With suggested actions, you as the bot developer would not need to consider this situation as suggested actions disappear from the chat window, as soon as the user proceeds the conversation. An example of suggested actions is shown in Figure 4-14.

*Figure 4-14.* *Suggested actions example*

## Adaptive Cards

The concept behind Adaptive Cards is to create cards once and let them be rendered natively by the channel the bot is deployed in. Whereas hero cards and other card types are usually part of the bot's code base, Adaptive Cards are authored using JSON. Therefore, the card itself is delivered to the channel in the JSON card format which makes it possible for the channel to render the card natively, using the same UI styles as for other parts within that channel. As of now, the channels outlined in Table 4-5 support

Adaptive Cards, which are by far not all Azure Bot Service channels. Therefore, you should first of all know which channels you want to deploy your bot to, before deciding which card type to choose.

---

**Note**  Not all Azure Bot Service channels support Adaptive Cards. To get a list of channels which support Adaptive Cards, please visit `https://docs.microsoft.com/en-us/adaptive-cards/resources/partners`.

---

***Table 4-5.*** *Adaptive Cards channel support (Microsoft, 2018)*

| Platform | Description | Version |
|---|---|---|
| Bot Framework Web Chat | Embeddable web chat control for the Microsoft Bot Framework. | 1.2.6 (Web Chat 4.9.2) |
| Outlook Actionable Messages | Attach an actionable message to email. | 1.0 |
| Microsoft Teams | Platform that combines workplace chat, meetings, and notes. | 1.2 |
| Cortana Skills | A virtual assistant for Windows 10. | 1.0 |
| Windows Timeline | A new way to resume past activities you started on this PC, other Windows PCs, and iOS/android devices. | 1.0 |
| Cisco WebEx Teams | Webex Teams helps speed up projects, build better relationships, and solve business challenges. | 1.2 |

As Adaptive Cards are basically JSON files or objects, it is simple to author cards. There is even an Adaptive Card designer available, which lets you design and author Adaptive Cards directly in your browser, without the need of writing JSON notations. Within Adaptive Cards, you can basically include text blocks, images, videos, columns, buttons, as well as input objects like text input fields, number or date input fields, or choice inputs.

---

**Note**  The Adaptive Card designer which can be used to author Adaptive Cards is available at `https://adaptivecards.io/designer/`.

---

Especially for developers which are new to Adaptive Cards, the designer is a good tool to get familiar with the available card elements and how to design a good-looking card. Moreover, the integrated preview mode offers you the possibility to view your authored Adaptive Card rendered natively in one of the support channels directly in the browser. Therefore, you can test the look and feel as well as the user experience of an Adaptive Card basically before developing your chatbot. The hero card shown in Figure 4-12 could also be created as an Adaptive Card like demonstrated in Figure 4-15.

*Figure 4-15.*  *Adaptive Cards example*

The card demonstrated in the preceding illustration is basically described in the following JSON object:

```
{
    "type": "AdaptiveCard",
    "body": [
        {
            "type": "Image",
            "url": "https://someUrl/someIcon.png",
            "size": "Large",
            "horizontalAlignment": "Center"
        },
```

```
{
    "type": "TextBlock",
    "size": "Large",
    "weight": "Bolder",
    "text": "Help?"
},
{
    "type": "TextBlock",
    "text": "Are you uncertain?",
    "wrap": true,
    "size": "Medium",
    "fontType": "Default",
    "isSubtle": true
},
{
    "type": "TextBlock",
    "text": "I can help you in any situation, where you may not
know how to continue the conversation! Just click one of the
buttons below to proceed",
    "wrap": true
},
{
    "type": "ActionSet",
    "actions": [
        {
            "type": "Action.Submit",
            "title": "Book a table",
            "data": "Book a table"
        },
        {
            "type": "Action.Submit",
            "title": "Manage booking",
            "data": "Manage booking"
        }
    ]
```

```
      }
   ],
   "$schema": "http://adaptivecards.io/schemas/adaptive-card.json",
   "version": "1.2"
}
```

But Adaptive Cards do not only offer the possibility to render static content and buttons. You can also create dynamic Adaptive Cards by toggling or showing some parts of the Adaptive Cards based on button taps or other actions as shown in Figure 4-16. This lets you build sophisticated UI elements which can be used throughout various channels without the need of developing or authoring the cards separately for each channel you want your bot to be accessible in. As Adaptive Cards are crucial for Bot Framework development, the next chapters will cover Adaptive Cards and their use cases in depth.

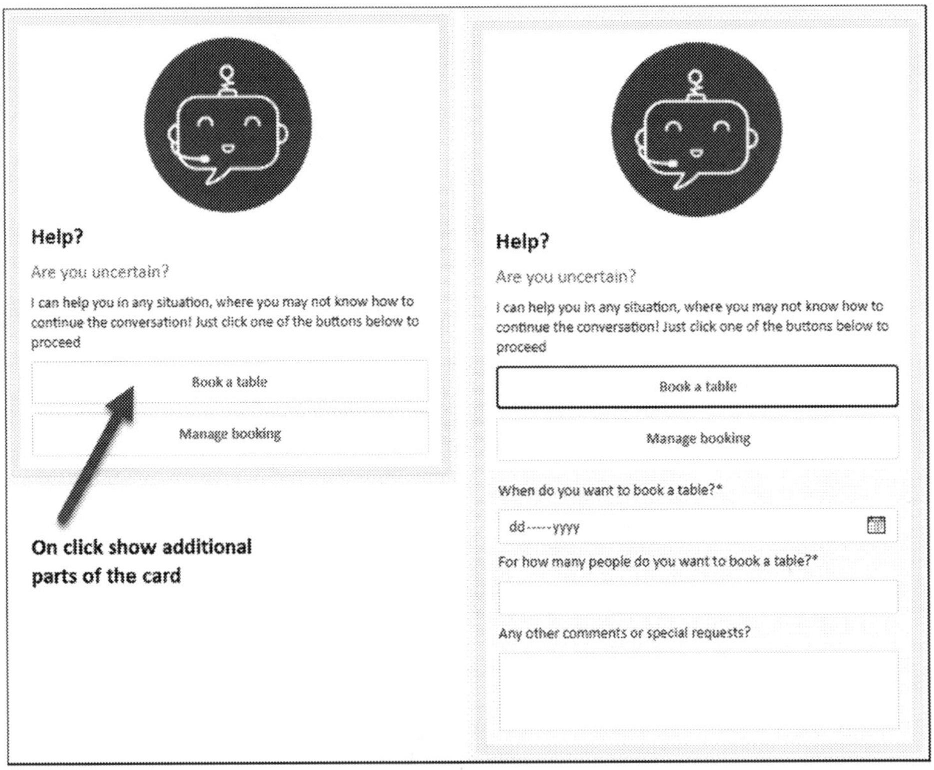

*Figure 4-16.* *Adaptive Cards toggle example*

# Summary

This chapter covered the design principles of a chatbot, which should be considered when building a successful chatbot. In the first part of this chapter, we examined the personality and branding topics of a chatbot. You learned why analyzing your chatbot's target audience is an essential task in the design phase and why a catchy greeting message and the navigation within a conversation are vital for every chatbot. In addition to that, you learned about the basic concepts for designing a conversation flow mainly focusing on the concept of adaptive dialogs. Moreover, we reviewed the user experience parts and available visual elements within a chatbot and why text-only chatbots are not always the best option.

In the next chapter, you will learn how to build a chatbot in a low-code approach using the Microsoft Bot Framework Composer, which is a new tool for building bots in a graphical user interface, which allows developers to add functionality via code as well.

# CHAPTER 5

# Building a Chatbot

In the last chapter, you have learned about designing a chatbot including the conversation flow and the user experience. This chapter will cover the topic of building a chatbot using a tool called "Bot Framework Composer," which is part of the Bot Framework ecosystem, for building chatbots using a graphical user interface instead of programming a chatbot using code. Moreover, the topics covered in previous chapters will be covered in detail within a real-life scenario for building a chatbot, such as using language understanding and QnA Maker, for enhancing natural language processing routines. In addition to that, this chapter will also describe the process of how to use custom code to include certain capabilities into a chatbot built with Composer, which is not available out of the box.

## Introduction to Bot Framework Composer

Basically, Bot Framework Composer is a tool with a graphical user interface, designed to author chatbots. Composer is targeting both developers and power users, offering many capabilities which are needed to build, test, and publish chatbots. Thus, Composer does also include the possibility to build up essential parts of a chatbot besides the core conversation flow, like the language understanding model, the language generation model, or QnA Maker knowledge bases. Therefore, it can be seen as a tool providing the most important features and capabilities for designing and developing chatbots, without the need of writing code.

Thus, Composer is designed to facilitate, for not only developers but also nondevelopers, the process of designing and building bots. Needless to say, the visual canvas capabilities accelerate the bot development life cycle, especially for people who do not regularly build a chatbot. In terms of using Composer, one has basically two options:

- Install Composer as a desktop application.

- Build Composer from source.

© Stephan Bisser 2021
S. Bisser, *Microsoft Conversational AI Platform for Developers*, https://doi.org/10.1007/978-1-4842-6837-7_5

Both scenarios have their advantages and disadvantages, but the overall look and feel as well as the features are the same in either option. Therefore, all examples which are referenced in this book are based on the desktop version of Bot Framework Composer. Composer consists of a graphical user interface as outlined in Figure 5-1, offering you a sophisticated authoring canvas, where dialogs, triggers, and actions can be built up.

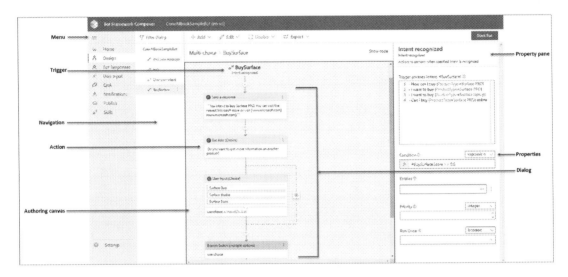

**Figure 5-1.** *Bot Framework Composer overview*

There are some differences when building a bot using Composer compared to the Bot Framework SDK, which a bot developer should keep in mind. First, Composer is based on the adaptive dialogs, described in Chapter 4, which is different to the dialogs library used in the Bot Framework SDK. At the same time, this is somehow one of the advantages of Composer, as adaptive dialogs are dialogs stored in JSON and are therefore reusable across many different bots or tools. In addition, another main advantage of using Composer over the SDK is that it combines the most common tools into one user interface. In the past, a bot developer would need to go to the LUIS web portal to manage all language understanding components. Furthermore, for managing QnA pairs, one had to go to the QnA Maker portal. And for the bot development itself, you would need to use Visual Studio or Visual Studio Code to develop the bot's logic. By introducing Composer, these experiences have been combined into one tool which saves developers a lot of time to set up and maintain the essential components for a bot.

---

**Note**    To download and install Bot Framework Composer, visit `http://aka.ms/bfcomposer`.

---

# Memory

In adaptive dialogs, the memory is basically referred to as the bot's mind as described in Chapter 4. Therefore, you can use the memory to store certain values which can be accessed at a later point within the conversation, like a user's name or address information. In Chapter 4 we learned that there are the following memory scopes, which offer different levels of accessibility and durability:

- User scope

- Conversation scope

- Dialog scope

- Turn scope

- Settings scope

- This scope

- Class scope

These scopes are used to store different values within the conversation, to reuse them later on. Each property you set in any given action or as a result to a question asked by the bot is stored in one of the abovementioned scopes. If you need to store data about a user or a conversation, you should use the user or conversation scope, as these scopes are durable throughout the bot's lifetime. The user scope is mostly used to store user information like the name, location, or other user properties and preferences. The conversation scope mainly targets conversation properties like the metadata of a conversation (e.g., start date, etc.) which are potentially shared by multiple users who have entered the same conversation (e.g., Microsoft Teams group chat). In case you need to store short-lived values, the scopes dialog and turn are better suited. The dialog scope's life cycle is basically bound to the dialog, which means that if the dialog ends, the properties defined in that scope are discarded. The turn scope is even more short-lived than the dialog scope, as properties defined in that scope are only retained within that turn, which is basically one single message handled by the bot.

To store properties in Composer, you can either use the "Set a property" action or set a property based on the "Ask a question" action. The first option basically is designed to set one single property at a given point within the dialog. There is also the option to set multiple properties at a time using the "Set properties" action. The "Set a property" action's behavior is shown in Figure 5-2. This action can essentially be used at any given step within the dialog to store information in one of the abovementioned scopes.

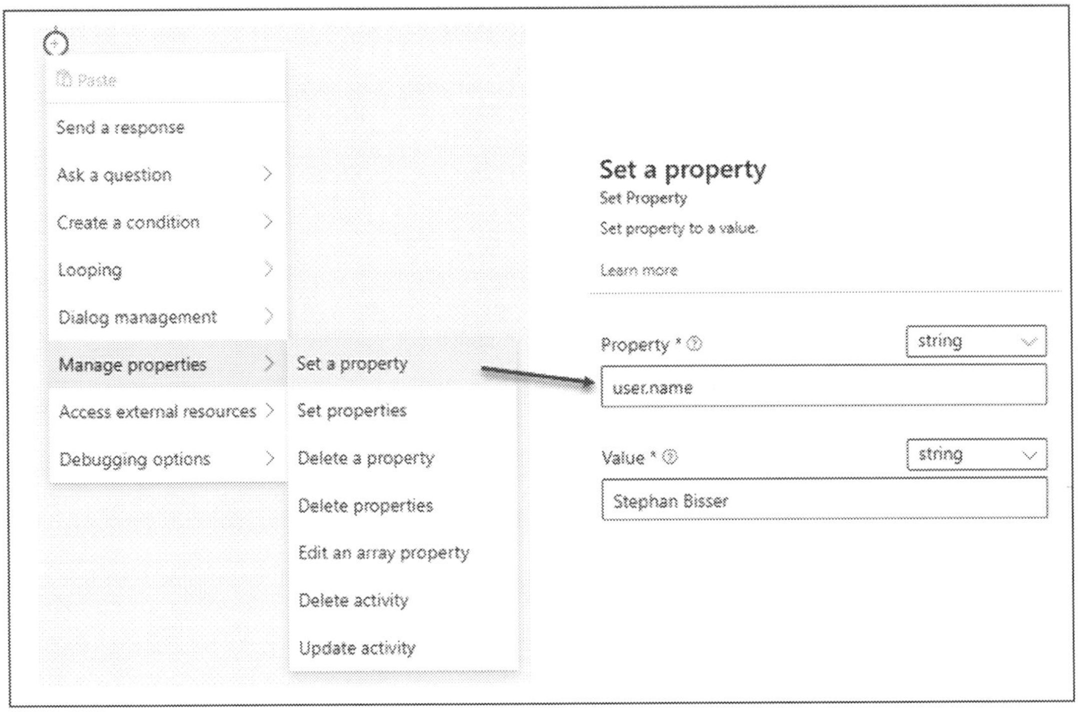

**Figure 5-2.**  *Set a property in Composer*

In the preceding example, the property's value is "hard-coded" and of type string. This however may not always be sufficient as you may want to generate values based on an operation. This concept in Composer is called "Expression." Expressions are basically used to calculate values based on computational functions. These expressions can include the following:

- Arithmetic operations like addition ("+"), subtraction ("-"), multiplication ("*"), or division ("/")

- Comparison operations like equals ("=="), not equals ("!="), greater than (">"), or less than ("<")

- Logical operations like and ("&&"), or ("||"), or not ("!")

Additionally, there are many prebuilt functions available which should be used to do operations on different object types like strings, collections, or dates as well as comparison or type-checking operations.

---

**Note**   To see the full list of prebuilt functions available, please conduct `https://docs.microsoft.com/en-us/azure/bot-service/adaptive-expressions/adaptive-expressions-prebuilt-functions`.

---

As Figure 5-3 demonstrates, you can also use *expressions* to calculate values using different prebuilt functions. As the following example shows, you can either set a property to a value hard-coded or use the *concat* function to calculate the property's value dynamically. This functionality also allows you to refer to existing properties within an expression by using the property's name within the expression, like *"=concat(user. first, ' ', user.last)."*

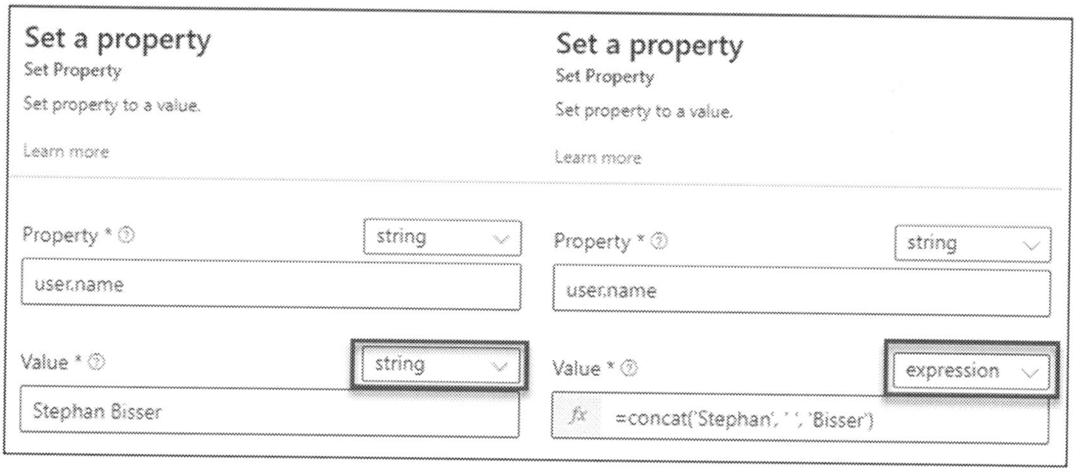

*Figure 5-3.  Set a property – comparison explicit type vs. expression*

To actually send out a message which contains a property's value, you simply need to use the ${scope.propertyName} notation as outlined in Figure 5-4. This notation basically is used in language generation to refer to properties stored in memory which need to be accessed for given actions.

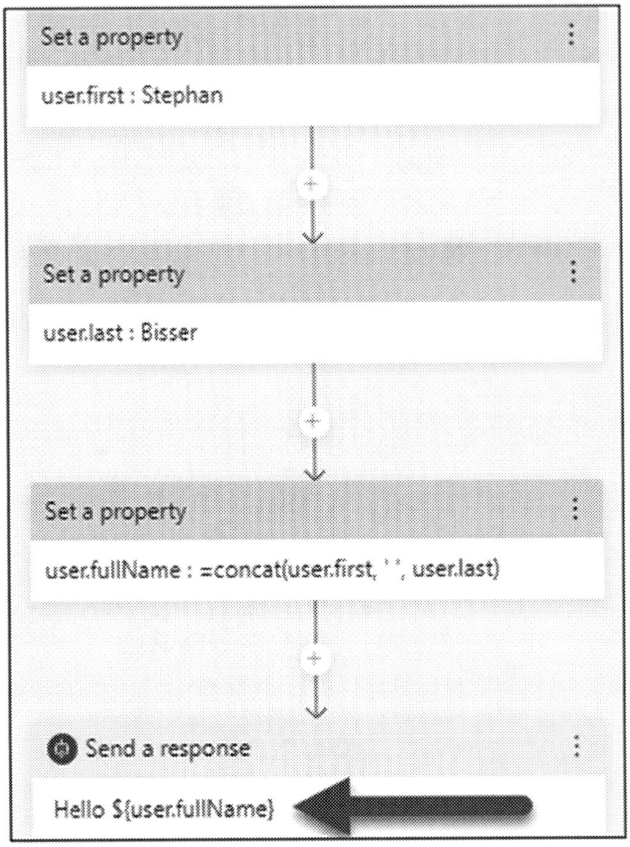

***Figure 5-4.***  *Send a message with property's value included*

# LU (Language Understanding)

As stated in earlier chapters, language understanding is a core functionality in nearly every conversational AI solution. Whenever you need to handle specific user request which go beyond simple QnA-based patterns, you need to implement some kind of language understanding to process user inputs accordingly. The main different between using LUIS in Bot Framework SDK projects compared to Composer is that you do not need to switch context between the IDE and the LUIS portal as Composer has an inline editor for authoring LUIS components, such as intents and entities.

The principles of maintaining a language model in Composer are basically the same as outlined in Chapter 3. Composer's inline editor lets you edit a .lu file which is used by Composer to manage the LUIS model automatically. These .lu files usually contain of intents, entities, and utterances. The notation used for each of these is outlined in Table 5-1.

***Table 5-1.***  *Composer .lu file notations*

| Type | Notation |
|------|----------|
| Intent | # IntentName |
| Utterance | - Some example utterance |
| Entity | {entityName=example} |

Therefore, the .lu file for the example intent of Chapter 3, GetWeather, could look like this:

```
# GetWeather
- could you tell me the weather forecast for redmond?
- how is the weather in new york?
- i would like to get some details on the weather in amsterdam please
- what is the weather like in seattle?
- what's the weather in berlin?
@ prebuilt geographyV2
```

You will notice that we add the prebuilt entity "geographyV2" to this intent, which automatically enables LUIS to detect and extract the location names from the user's utterances. The "BookTable" intent used within Composer would look like the following:

```
# BookTable
- book a table at the {@locationName=redmond steak house} and grill please.
- can you reserve a table for 4 people at the {@locationName=famous sushi
  bar} tomorrow?
- could you reserve a table at {@locationName=tom's diner} tomorrow?
- please reserve a spot for 5 at {@locationName=jamie's kitchen} on 23rd of
  november.
```

```
- reserve a table for saturday 3 p.m. at {@locationName=hard rock café} for
six people please
@ ml locationName
@ prebuilt datetimeV2
@ prebuilt number
```

The use of these two .lu files within Composer will be described in detail in the following parts within this chapter, along with some further information on how to handle language understanding accordingly.

# LG (Language Generation)

As language understanding provides bot developers an easy way of adding natural language processing features to a chatbot, language generation is mainly used to give your chatbot a tone of voice to make it look more intelligent. Language generation is therefore intended to give developers the flexibility of adding variations to the chatbot's messages. This offers the possibility of not only detaching the business logic from the presentation layer but also using a consistent concept for managing and maintaining language generation parts within a chatbot solution, by using .lg files. The .lg file format is quite simple and somewhat similar to the .lu file format described earlier in this chapter and is described in Table 5-2:

```
> Greeting template with 3 variations.
# Greeting
- Hi
- Hello
- Howdy
```

*Table 5-2.  Composer. lg file format*

| Component | Notation |
|---|---|
| Comment | > This is a comment |
| Template name | # TemplateName |
| Variation | - This is a bot reply |

- .NET Core SDK 3.1 and higher

  - `https://dotnet.microsoft.com/download/dotnet-core/3.1`

- Composer

  - `https://aka.ms/bfcomposer`

# Templates and Samples

After you have installed the three tools mentioned in the preceding text, you can start using Composer to author your first chatbot. To support people who are starting to use the Bot Framework SDK and Composer, there are many templates and examples available which act as a boilerplate for your next project. The predefined samples are outlined in Table 5-3 along with a short description about each sample.

***Table 5-3.*** *Bot Framework Composer samples (Microsoft, 2020)*

| Sample | Description |
| --- | --- |
| **Echo bot** | A bot that echoes whatever message the user enters. |
| **Empty bot** | A basic bot that is ready for your creativity. |
| **Simple to-do** | A sample bot that shows how to use Regex recognizer to define intents and allows you to add, list, and remove items. |
| **To-do with LUIS** | A sample bot that shows how to use LUIS recognizer to define intents and allows you to add, list, and remove items. A LUIS authoring key is required to run this sample. |
| **Asking questions** | A sample bot that shows how to prompt user for different types of input. |
| **Controlling Conversation flow** | A sample bot that shows how to use branching actions to control a conversation flow. |
| **Dialog actions** | A sample bot that shows how to use actions in Composer (does not include **Ask a question** actions already covered in the **Asking Questions** example). |

*(continued)*

***Table 5-3.*** (*continued*)

| Sample | Description |
|---|---|
| **Interruptions** | A sample bot that shows how to handle interruptions in a conversation flow. A LUIS authoring key are required to run this sample. |
| **QnA Maker and LUIS** | A sample bot that shows how to use both QnA Maker and LUIS. A LUIS authoring key and a QnA Knowledge Base are required to run this sample. |
| **QnA sample** | A sample bot that is provisioned to enable users to create QnA Maker knowledge base in Composer. |
| **Responding with cards** | A sample bot that shows how to send different cards using language generation. |
| **Responding with text** | A sample bot that shows how to send different text messages to users using language generation. |

# Create an Echo Bot

The first chatbot we will create is based on the "Echo Bot" sample, which basically is a simple bot echoing back everything the user types into the chat. To create a new bot, you need to choose "+ New" from Composer's main screen and then choose "Create from template" and select "Echo Bot" as demonstrated in Figure 5-5.

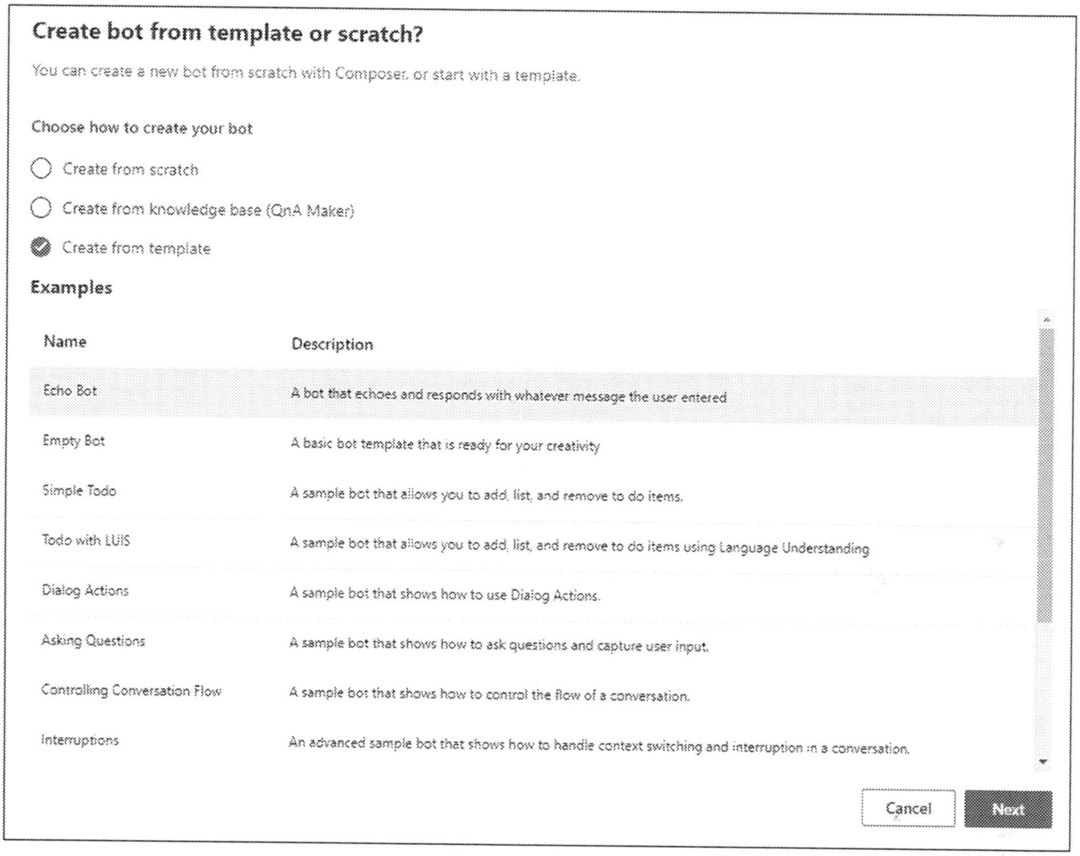

**Create bot from template or scratch?**

You can create a new bot from scratch with Composer, or start with a template.

Choose how to create your bot

○ Create from scratch

○ Create from knowledge base (QnA Maker)

◉ Create from template

**Examples**

| Name | Description |
|---|---|
| Echo Bot | A bot that echoes and responds with whatever message the user entered |
| Empty Bot | A basic bot template that is ready for your creativity |
| Simple Todo | A sample bot that allows you to add, list, and remove to do items. |
| Todo with LUIS | A sample bot that allows you to add, list, and remove to do items using Language Understanding |
| Dialog Actions | A sample bot that shows how to use Dialog Actions. |
| Asking Questions | A sample bot that shows how to ask questions and capture user input. |
| Controlling Conversation Flow | A sample bot that shows how to control the flow of a conversation. |
| Interruptions | An advanced sample bot that shows how to handle context switching and interruption in a conversation. |

Cancel    Next

***Figure 5-5.***   *Create first chatbot 01*

After selecting the template, you need to give your chatbot a name and choose the location where you want Composer to store the chatbot project in. In my case the chatbot's name is "Diele," and I also enter an optional description, to describe the main objectives, as illustrated in Figure 5-6.

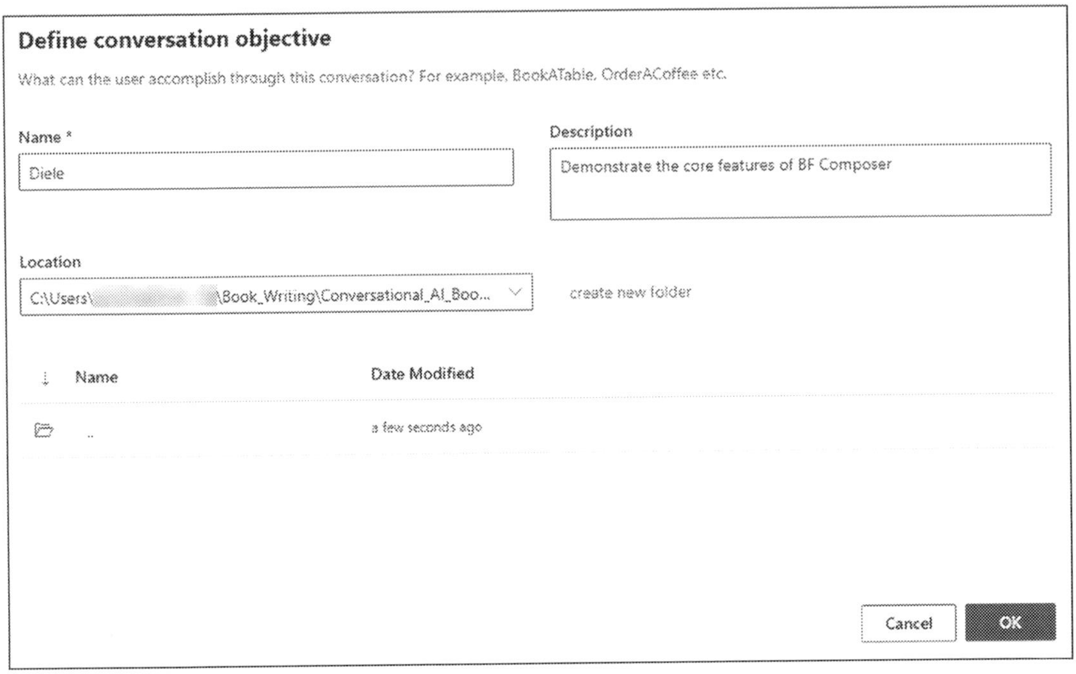

**Define conversation objective**

What can the user accomplish through this conversation? For example, BookATable, OrderACoffee etc.

Name *

Diele

Description

Demonstrate the core features of BF Composer

Location

C:\Users\ ⬛⬛⬛⬛⬛⬛ \Book_Writing\Conversational_AI_Boo...   ∨          create new folder

|   | Name | Date Modified |
|---|------|---------------|
| 🗁 | .. | a few seconds ago |

Cancel      OK

***Figure 5-6.*** *Create first chatbot 02*

After creating your first bot, Composer will show the newly created bot along with its "Greeting" trigger, as shown in Figure 5-7. As we chose the "Echo Bot" template, our bot basically consists of one dialog with two triggers. The first trigger is the "Greeting" trigger which gets triggered on every "ConversationUpdate" activity, whereas the second trigger is called the "Unknown intent" trigger which gets triggered whenever the bot is capable of recognizing the correct intent.

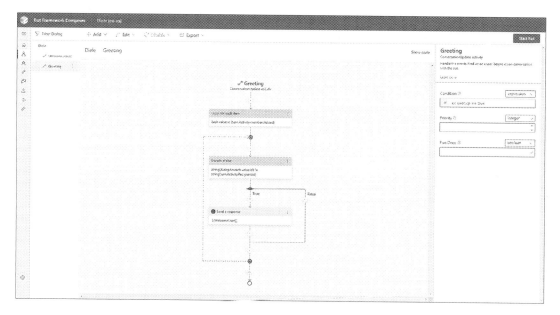

***Figure 5-7.*** *Create first chatbot 03 – greeting trigger*

The business logic within the "Unknown intent" trigger is very basic as outlined in Figure 5-8. This trigger mainly consists of the action for echoing back what the user wrote. As you can see in this example, the last message from the user to the bot is always stored in the property "turn.activity.text" within the turn scope, which is only valid within the given turn.

***Figure 5-8.*** *Create first chatbot 04 – unknown intent trigger*

Now to test your bot the very first time, you can press the "Start Bot" button from the top right corner in Composer. This will basically run the bot on your local machine and prepare everything which is needed to test your chatbot locally. After the bot has been started, you can simply click the "Test in Emulator" button next to the "Restart Bot" button from the top right corner to open your chatbot in the Bot Framework Emulator, as illustrated in Figure 5-9.

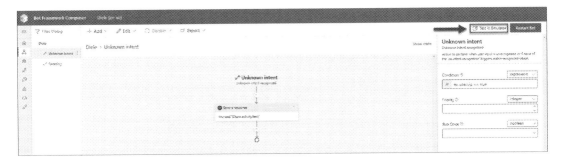

***Figure 5-9.*** *Create first chatbot 05 – test bot*

After clicking the "Test in Emulator" button, the Bot Framework Emulator should be opened, and the conversation with the bot should be established. You will notice that the bot will send you a greeting message of "Welcome to the EchoBot sample" as it is defined in the "Greeting" trigger within the "WelcomeUser" language generation template. Now if you type a message into the send box and send it to the chatbot, you will see that the bot will respond with "You said:" and the message you sent. So it basically uses the "Send a response" action from the "Unknown intent" trigger to send out the message containing these two parts, as shown in Figure 5-10.

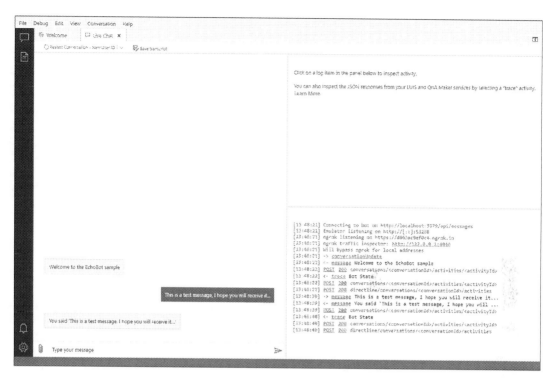

***Figure 5-10.*** *Create first chatbot 05 – test bot in emulator*

Now as you can see from the Composer user interface, there are a lot of options to choose from when authoring dialogs, as shown in Figure 5-11. There is the possibility to send a response using language generation, which has been described earlier and which was used to send the greeting message as well as the echo message.

*Figure 5-11.* *Composer authoring options*

You can also include questions in your dialogs, to query specific topics directly within the conversation. There are a lot of prebuilt prompt options available, as outlined in Figure 5-12. Most of these have been discussed in Chapter 2 already, which is why they will not be covered in detail in this chapter. Using these prompts, you can ask a question in a specific part of the conversation and expect a certain answer type, like a text, a number, or even an authentication event using OAuth. This can be extremely helpful to store certain values into properties, which can then be used later on in the conversation.

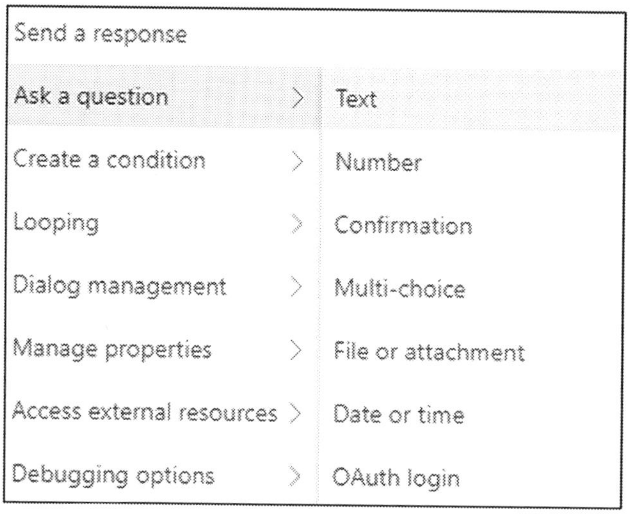

| Send a response | | |
|---|---|---|
| Ask a question | > | Text |
| Create a condition | > | Number |
| Looping | > | Confirmation |
| Dialog management | > | Multi-choice |
| Manage properties | > | File or attachment |
| Access external resources | > | Date or time |
| Debugging options | > | OAuth login |

**Figure 5-12.** *Composer authoring options – ask a question*

A very basic example of how to use the prompt feature would be to ask the user for his or her name within the greeting trigger. So after each loop, we simply add a new action of type "Ask a question – Text" and insert some prompt texts into the "Bot Asks" section on the right hand side as shown in Figure 5-13. These prompts could be something like the following example prompts:

- What is your name?

- Could you tell me your name please?

- What's your name please?

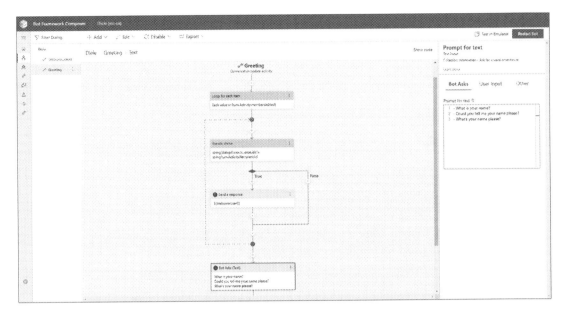

**Figure 5-13.** *Create a text prompt 01*

Now we need to switch to the "User Input" section to enter the name of the property which should be used to store the name in. As we might need the name of the user throughout the whole conversation, we basically have the option to use the "user" scope or the "conversation" scope for this. As the name is tightly bound to the user, the "user" scope is probably better suited for this property, so we enter "user.name" into the "Property" field as outlined in Figure 5-14.

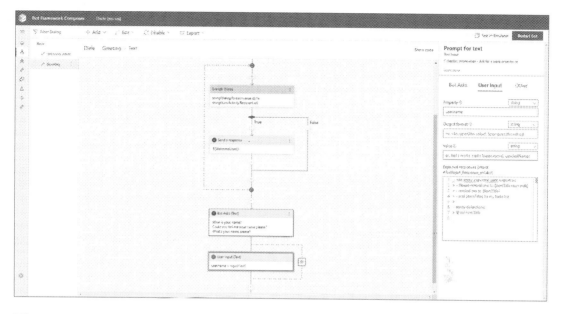

**Figure 5-14.**  *Create a text prompt 02*

The last thing we need to do is to send a personalized message to the user upon entering his or her name. Therefore, we add another action after the "Bot Asks (Text)" action of type "Send a response." Within the language generation, you can basically enter any text you want, but if you want to include the person's name in the message, you will need to use the notation of *${user.name}* to refer to the value of the property "user.name" in memory, as illustrated in Figure 5-15.

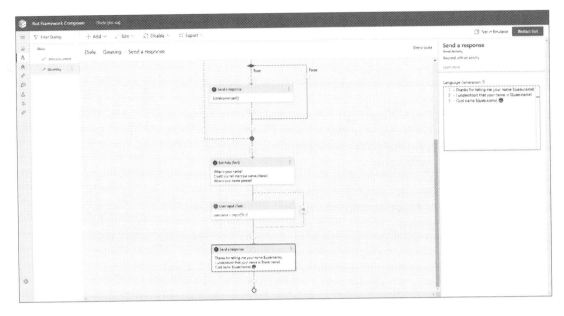

***Figure 5-15.*** *Send a response including a property*

To validate the new conversation behavior, we can simply click "Restart Bot" from the top right corner of the Composer screen, to restart the bot. After the restart has been finished, we can switch to the Bot Framework Emulator and click "Restart conversation" to restart the conversation from the beginning. If everything worked out correctly, you should see two messages from the bot at the beginning, the first one is the default greeting message, and the second one should be the question around your name. After entering your name, you should see the response of the bot including your name. One thing you may notice when restarting the conversation again and answering the question again is that the language generation engine should pick different variations almost each time, as shown in Figure 5-16. This basically makes the bot more personalized and human-alike as the conversation manner is different than if the bot would always use the exact same phrasing for the same responses.

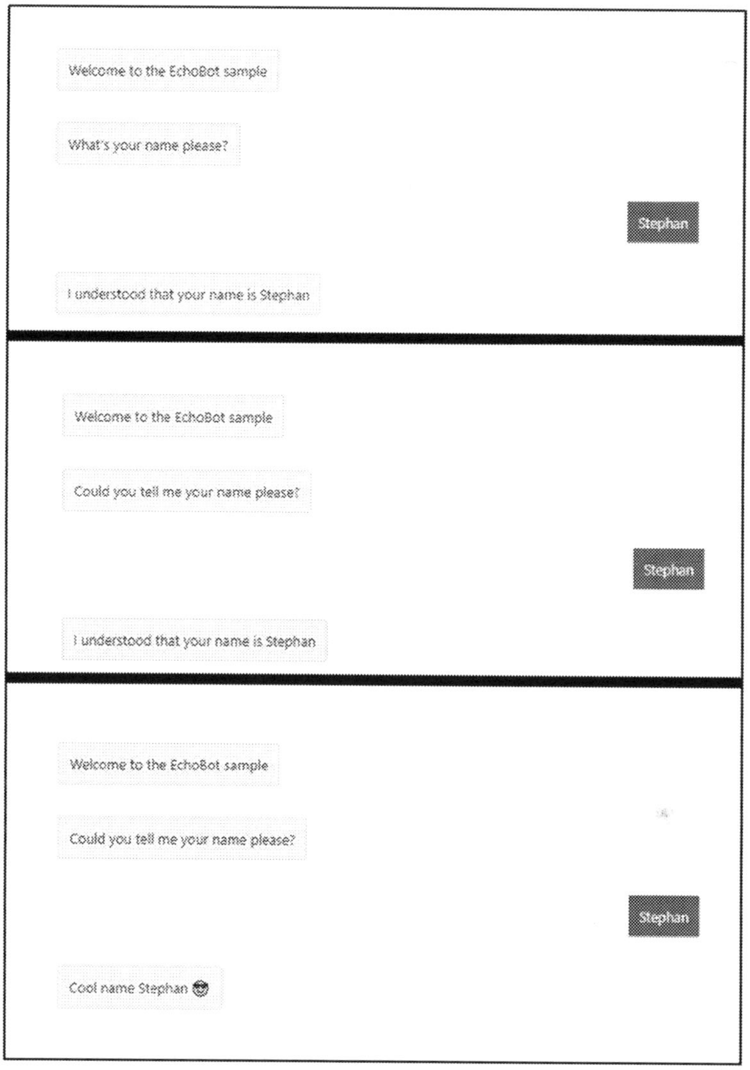

***Figure 5-16.*** *Test bot and question behavior*

# Enhance Chatbot with LUIS and QnA Maker

Now that you have covered a very basic use case using Composer, the next step would be to cover the dialog tree example mentioned in Chapter 4. This dialog tree is also represented in Figure 5-17, although the overview has been enhanced by adding the "QnADialog" as well to include questions which need to be answered by the bot which are not necessary bound to one specific action within the other dialogs. The following illustration basically covers all triggers and dialogs which will be built during this

199

chapter. The primary goal is to have a chatbot up and running which is able to cover the booking of a restaurant table as well as the management of existing bookings. In addition, the bot should be able to give information about the weather forecast using a third-party API, which will be part of the next subchapter. Furthermore, the chatbot should be able to answer questions around the restaurant or the booking process in general, which will be served via a QnA Maker knowledge base.

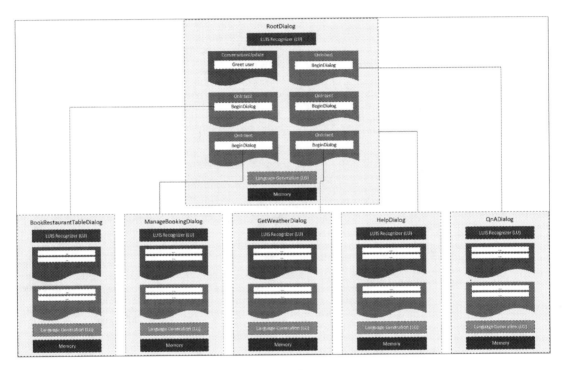

***Figure 5-17.*** *Overview of the necessary dialogs*

Before going ahead with authoring the dialogs, the first thing you should do is to modify the greeting part of the conversation. As we learned in Chapter 4, the greeting is essential to the success or failure of the bot as it will leave either a good or bad first impression at your users. Therefore, you should probably use a greeting in the form of a card to let the bot introduce itself proactively instead of sending a plain text message as it is now. For the sample bot discussed in this chapter, the bot will use an Adaptive Card as a greeting message, which eventually is demonstrated in Figure 5-18.

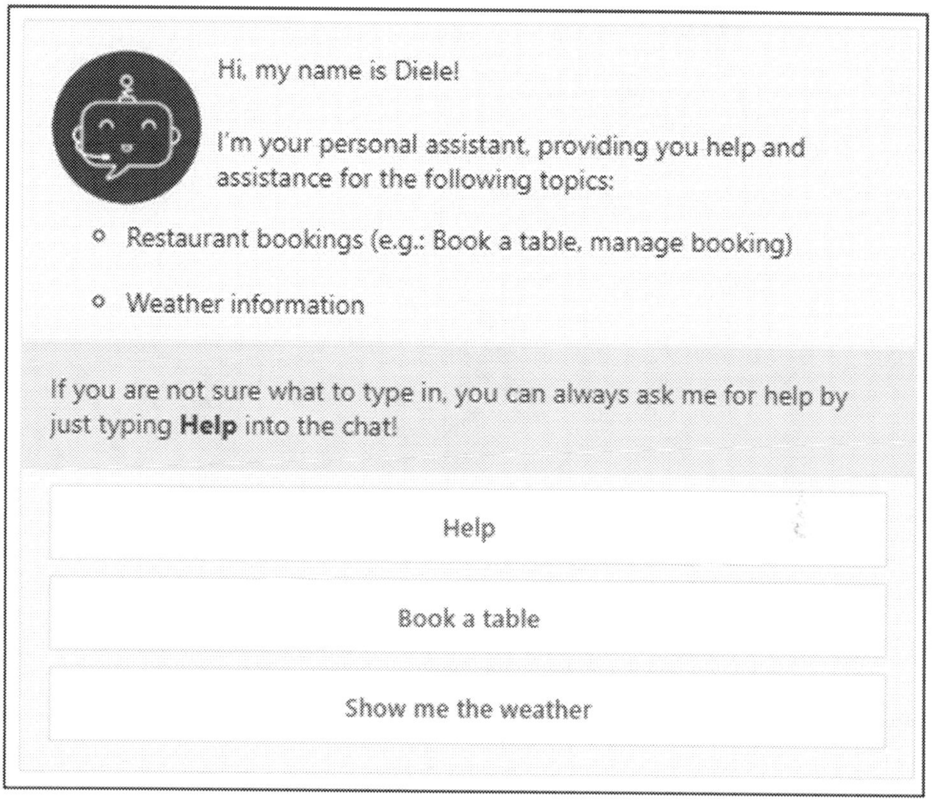

**Figure 5-18.** *Greeting Adaptive Card example*

To include an Adaptive Card in a message within Composer, you need to add the card's JSON payload to the bot responses section in the respective dialog. The formatting of the method is outlined in Figure 5-19. The Adaptive Card JSON payload should be added to a function which is called "GreetingCardJson" in this example.

**Figure 5-19.** *Insert an Adaptive Card into the bot responses*

The card itself can then be created in a second function as shown in the following code snippet, which is called *"GreetingCard."* This function basically converts the payload from the first function which is basically a string to a JSON payload. This payload is then used as an attachment within a new bot activity:

```
# GreetingCard
[Activity
    Attachments = ${json(GreetingCardJson())}
]
```

The greeting card can then be sent as a response using the *"GreetingCard"* function in a notation of *"${GreetingCard()}"*, as shown in Figure 5-20. This basically calls the "GreetingCard" function, which generates a new activity with an attachment which gets returned by the *"GreetingCardJson"* function.

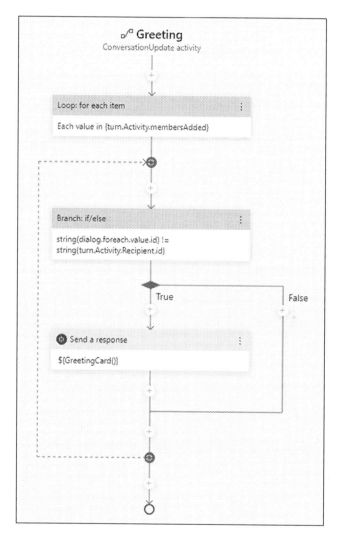

**Figure 5-20.** *Greeting dialog*

Now that the greeting is implemented accordingly, the next step would be to enhance your chatbot with the use cases mentioned in the preceding text. Therefore, you need to add a couple of triggers to your main dialog. To add a new trigger, simply click the "+ Add" button from the top of Composer, select "Add new trigger …," and insert the name as well as the trigger phrases as outlined in Figure 5-21.

203

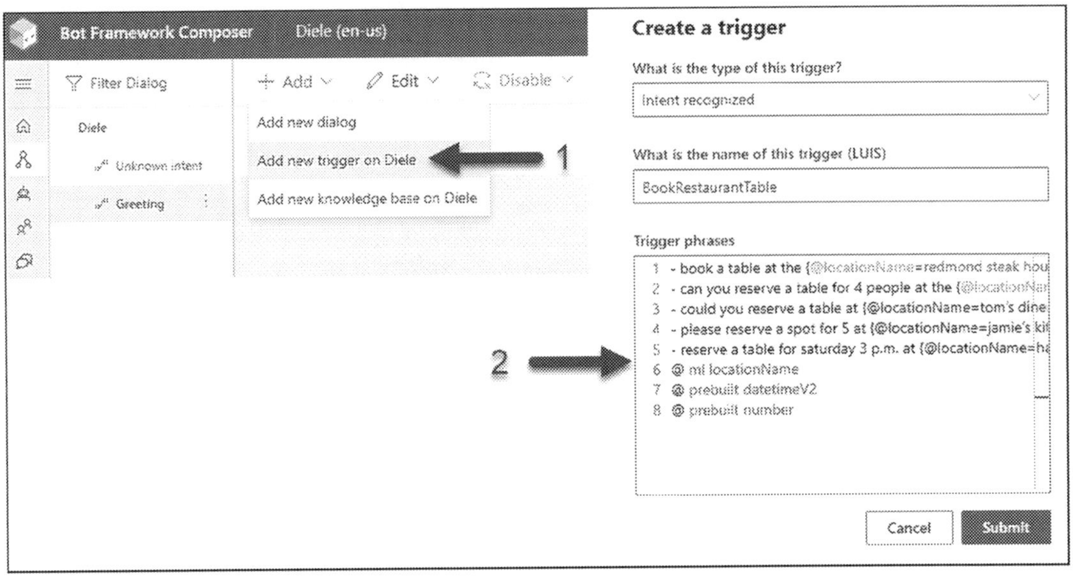

*Figure 5-21.  Add a new trigger in Composer*

The following table describes which triggers need to be added and which type of triggers these are. Furthermore, the trigger phrases, which can be used as sample utterances, are also present in Table 5-4.

*Table 5-4.  Sample chatbot triggers*

| Name | Type | Trigger Phrases |
|------|------|-----------------|
| **BookRestaurantTable** | Intent recognized | - book a table at the {locationName=redmond steak house and grill} please<br>- can you reserve a table for 4 people at the {locationName=famous sushi bar} tomorrow?<br>- could you reserve a table at {locationName=tom's diner} tomorrow?<br>- please reserve a spot for 5 at {locationName=jamie's kitchen} on 23rd of november<br>- reserve a table for saturday 3 p.m. at {locationName=hard rock café} for six people please<br>- I want to book a restaurant table please<br>@ ml locationName<br>@ prebuilt datetimeV2<br>@ prebuilt number |

*(continued)*

***Table 5-4.*** (*continued*)

| Name | Type | Trigger Phrases |
|------|------|-----------------|
| **ManageBooking** | Intent recognized | - i want to change my booking please<br>- can I adapt my booking with numer {bookingNumber=123abc}<br>- I need to make a change in my reservation {bookingNumber=456def}<br>- could you maybe modify my booking with the reference {bookingNumber=789ghi}<br>- please let me change one of my reservations<br>@ ml bookingNumber |
| **GetWeather** | Intent recognized | - please tell me the weather<br>- what is the weather like in {city=Seattle}<br>- could you tell me the weather forecast for {city=redmond}?<br>- how is the weather in {city=new york}?<br>- what's the weather in {city=berlin}?<br>@ ml city |
| **Help** | Intent recognized | - Help<br>- I need help<br>- could you assist me please<br>- I need assistance<br>- Please help me |
| **QnA** | QnA Intent recognized | *(No need to add trigger phrases)* |

The next step would be to set a specific condition when the triggers should be executed. As LUIS recognizers are used for the abovementioned triggers, the condition should be bound to the confidence score of the LUIS recognizer. In general, a confidence score of 0.8 and above is good and basically indicates a good confidence level within the language understanding part of your bot. To set a condition for a specific trigger, open up the trigger in Composer, and set the condition to *#TriggerName.Score >= 0.8*, as shown in Figure 5-22.

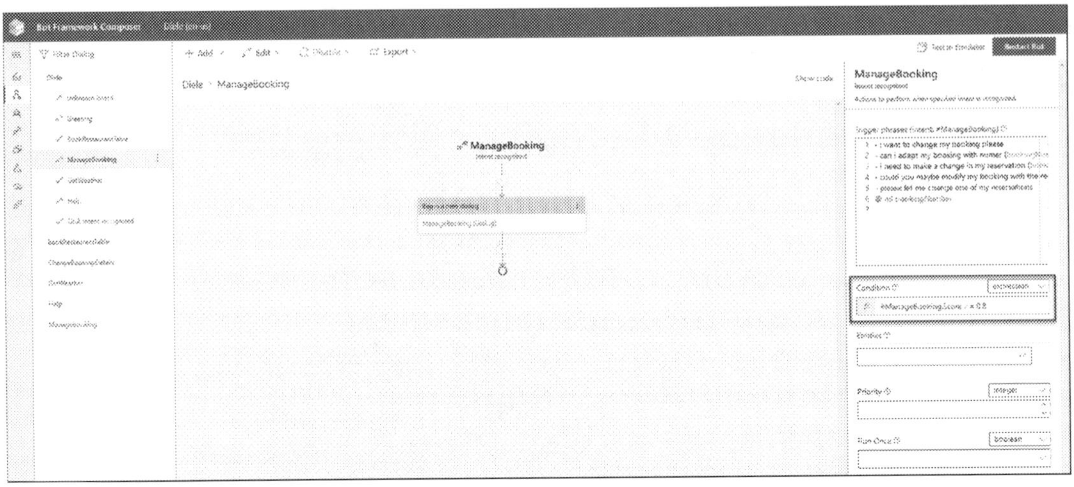

**Figure 5-22.** *Add a condition to a trigger*

Now that all triggers have been created, you can go on and start a new dialog for each
of the abovementioned triggers, to handle the dialog's actions separately. Therefore,
within each of the recently created triggers, add a new action of type "Begin a new
dialog," and create a new dialog for each trigger, as shown in Figure 5-23.

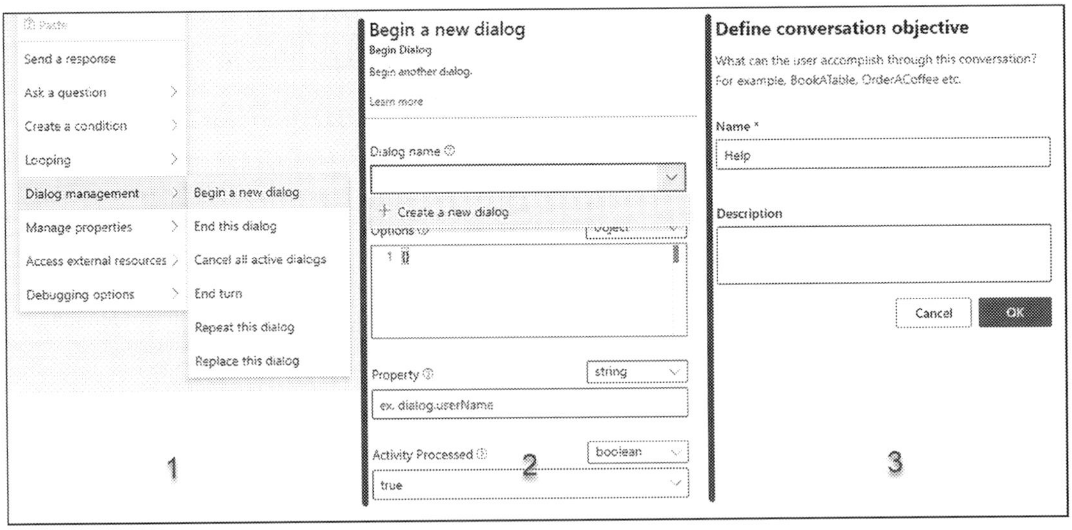

**Figure 5-23.** *Begin new dialog within each trigger*

# BookRestaurantTable Dialog

Now you can begin to author the dialogs to include some business logic in the conversation. The "BookRestaurantTable" dialog should simply gather all necessary information to create a new booking reference for a specific user. Therefore, it is necessary to get the information about the location, the date, and the number of people which the user would like to book a table for. All three of these properties are marked as entities within the language understanding model, which should be extracted automatically. Therefore, the easiest way to capture these entities is to use the action "Set properties" as a first step within the "BookRestaurantTable" dialog to store the information extracted by LUIS in separate properties within the dialog scope, as demonstrated in Figure 5-24. The important step to note here is the usage of the "coalesce" prebuilt function. This function basically returns the first nonnull value from the list of parameters, which are in this case the entity property, which is annotated with an "@" (e.g., "@locationName") at the beginning and the dialog's property which is prefixed with a "$" (e.g., "$locationName").

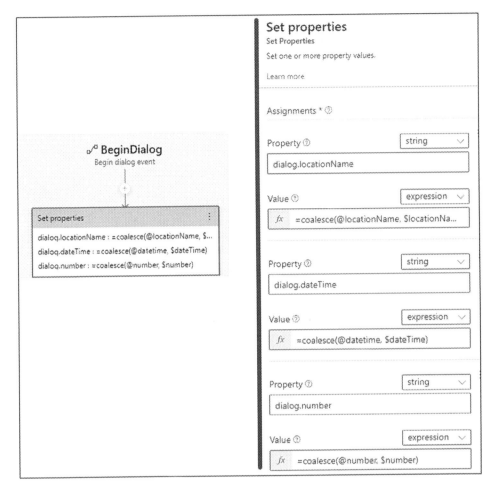

***Figure 5-24.*** *BookRestaurantTable dialog – set properties*

The next step is to use the "Ask a question" action to ask for the missing properties to fill, before going to the next step of the conversation. Therefore, you can add three actions to ask for the location's name, the date for the reservation, as well as the number of people to book a table for, as outlined in Figure 5-25.

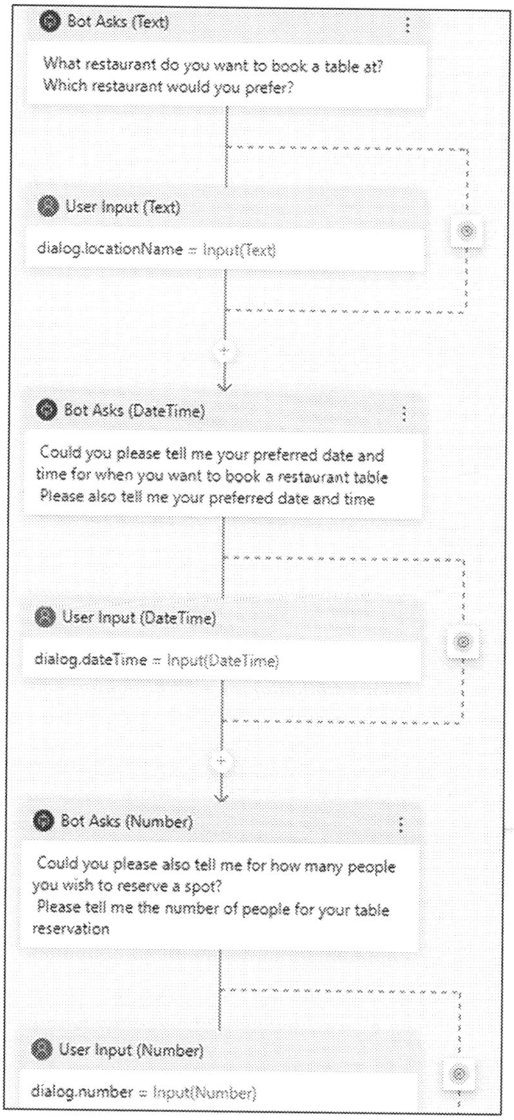

***Figure 5-25.*** *BookRestaurantTable – ask a question action*

There is a helpful feature in Composer, which should be used in occasions like these, which prevents the bot from asking the question even if the property is already filled, which is called "Always prompt." If this attribute is set to true, the question will be asked in any case, even if the property, which should hold the answer's value, is not null or empty. As the first action in this dialog stores the extracted entities into properties, it might happen that LUIS already extracted the values automatically from the initial utterance. Therefore, the bot already knows the property's value and does not need to ask again for the location's name, the date, or the number of people and can skip those questions to prevent a bad user experience. Thus, this setting should be set to false in this case as shown in Figure 5-26.

*Figure 5-26.* *BookRestaurantTable – always prompt feature*

The next step within the dialog would be to ask for a confirmation before going ahead and add the booking to the system. This can be done using the "Ask a question" action of type "Confirmation," as shown in Figure 5-27. As you can see in the following illustration, the user's input is stored in a property called "dialog.confirm," which will be used in a later step.

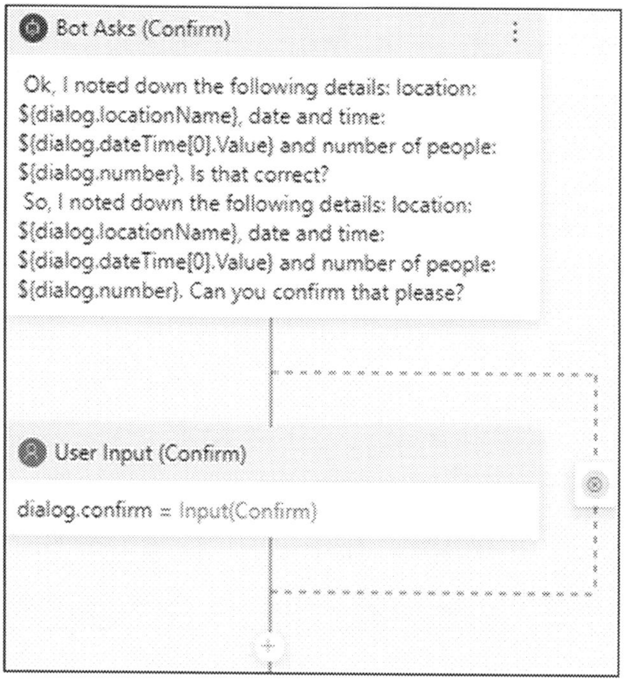

**Figure 5-27.** *BookRestaurantTable dialog – ask for confirmation*

The last step within the dialog is to check the result of the confirmation using a "Branch: If/else" action as outlined in Figure 5-28. If the user confirmed that the details are correct, then a response will be sent out, stating that the booking will be done by the bot and a confirmation email will be sent out as well. In addition, the booking details will be stored in an array called "user.bookings" where all bookings from a user will be stored in for later reference. If the user did not confirm the booking details, then the bot asks back if the user wants to change anything regarding the booking. If this is the case, then the dialog will be simply repeated, whereas all properties will be deleted beforehand, so that the bot will ask the user all three questions again to gather the correct information. If the user does not want to change anything, the dialog will be ended which will result in the fact that the dialog which called this dialog will be resumed.

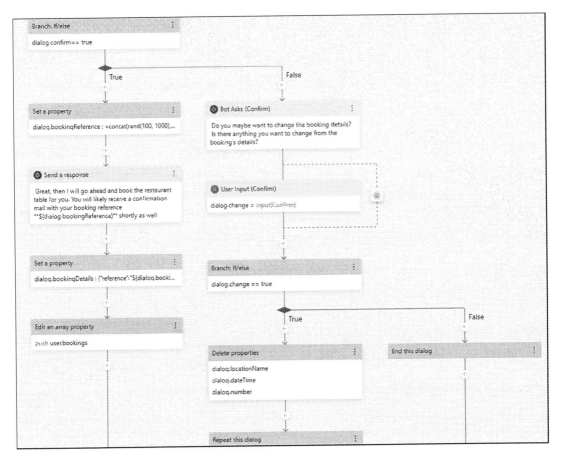

**Figure 5-28.** *BookRestaurantTable dialog – confirmation options*

The last action within this dialog before it ends would be to let the user know that if there is anything else to fulfill, the bot is likely there to help. Therefore, you should add an action of type "Send a response" as shown in Figure 5-29 to send out a message of type "What else can I help you with?" followed by the "End this dialog" action. This will make sure that the user will receive a message at the end of this dialog to let him or her know that the bot is expecting another question or input, if the user still has a question. As the dialog will be ended by the "End this dialog" action, the parent dialog will be resumed which assures that the LUIS recognizer will be used to determine the intent of the new utterance correctly to start the according dialog.

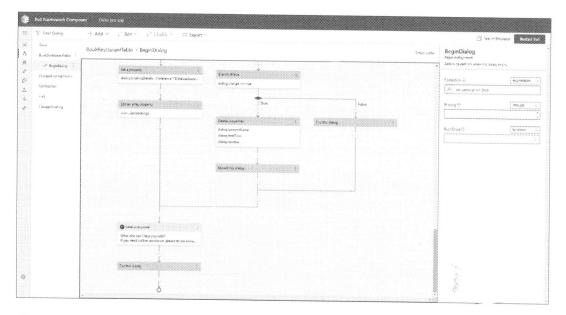

***Figure 5-29.*** *BookRestaurantTable dialog – end dialog action*

The complete dialog is outlined in Figure 5-30, with all used actions and steps, which have been discussed in the section before to get an overview of the complete dialog/ trigger.

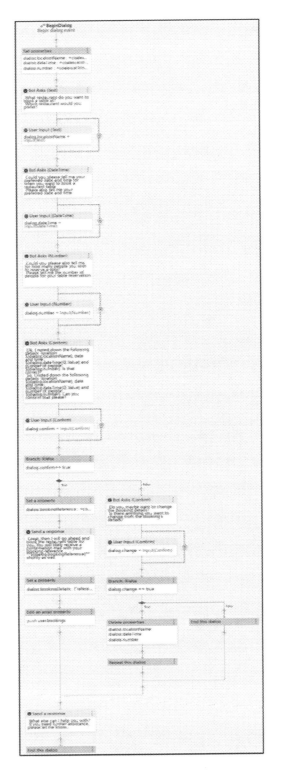

**Figure 5-30.** *BookRestaurantTable dialog – overview*

**Note**  In a real-world scenario, you would probably call a third-party service, which is used to manage bookings and reservations. Therefore, the bot is used to gather all necessary information from the user and then call an API for instance to create the reservation in the booking system accordingly.

# ManageBooking Dialog

As the bot is now capable of handling restaurant reservations, the next step would be to handle inquiries for users who seek help in terms of changing existing reservations. Therefore, the "ManageBooking" dialog should be used to change specific details within an existing booking reservation. The first step in this dialog would be to set the booking reference property, similar to the first step in the "BookRestaurantTable" dialog, as shown in Figure 5-31. This action is used to set the value of the property "dialog.bookingReference" to the extracted entity value received from the LUIS recognizer.

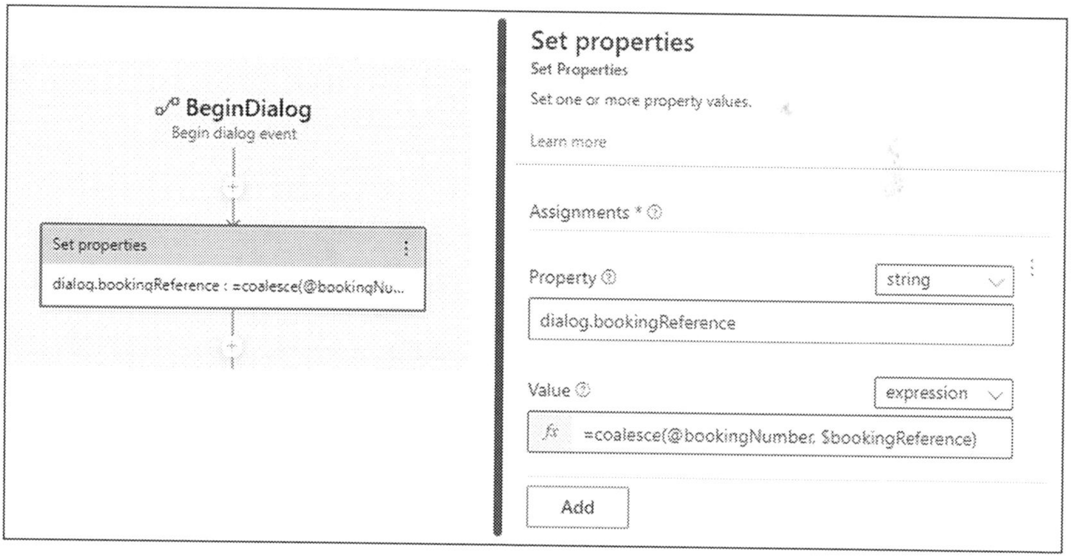

*Figure 5-31. ManageBooking dialog – set properties*

If the LUIS recognizer was not able to extract the booking reference number, then the next action would be to ask the user for the booking reference as demonstrated in Figure 5-32. Similar to the previous dialog, this action will only be executed if the property "dialog.bookingReference" is null or empty. Therefore, this question will only be sent to the user, if the LUIS recognizer has not successfully extracted the booking reference entity.

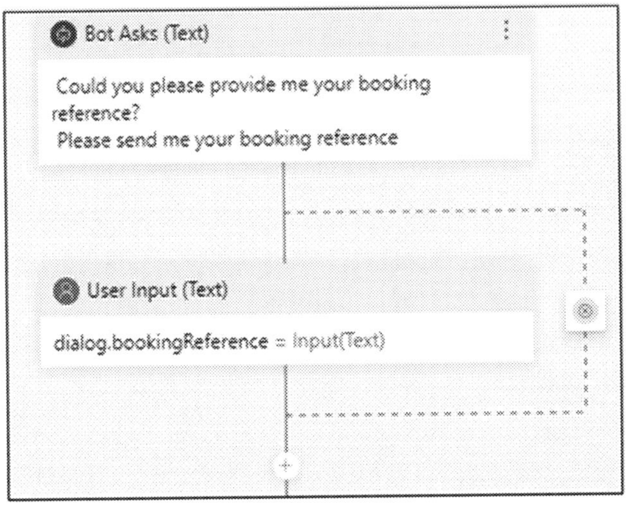

***Figure 5-32.*** *ManageBooking dialog – ask for booking reference*

Now that the bot received the booking reference number, the next step is to compare this number with the booking reference numbers stored in the property "user.bookings," as displayed in Figure 5-33. If a match has been found, the booking will be removed from the user's bookings, and an additional dialog "ChangeBookingDetails" will be executed to modify the properties of the booking separately.

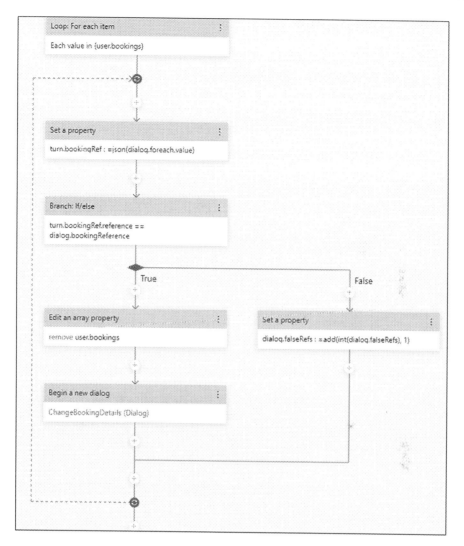

**Figure 5-33.** *ManageBooking dialog – call ChangeBookingDetails for matching booking reference*

In Figure 5-34, the starting action of the "ChangeBookingDetails" is displayed. In this dialog, the user has to select which property of the booking should be modified. Therefore, the user has the option to select a choice from the multichoice question action received by the bot.

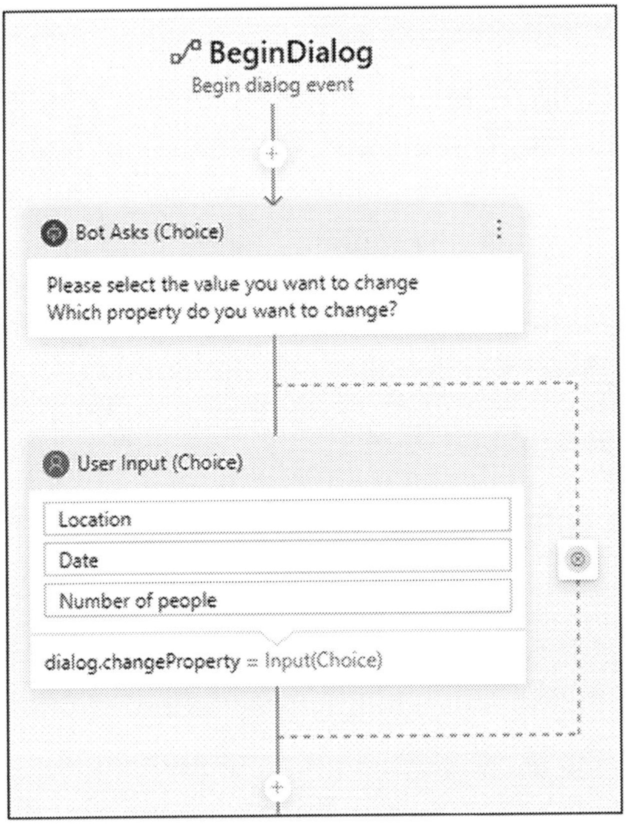

**Figure 5-34.** *ChangeBookingDetails dialog – select property to change*

After the user has selected one of the offered choices, the "Branch: Switch (multiple options)" action is used to determine which option has been selected by the user as shown in Figure 5-35. Depending which option the user chose, the corresponding property will be set, by asking the user about the new or desired value with a text prompt.

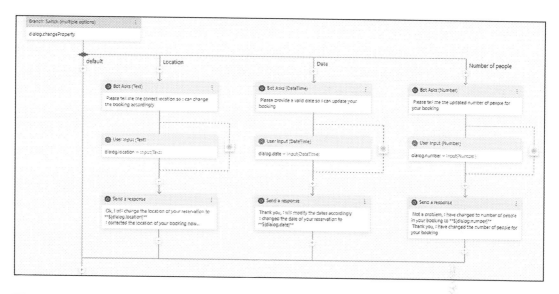

**Figure 5-35.** *ChangeBookingDetails dialog – change the specified property accordingly*

As outlined in Figure 5-36, the new "dialog.bookingDetails" property will be created, which holds the new properties for the user's bookings. In addition, the action "Edit an array property" is used to push the property "dialog.bookingDetails" into the array "user. bookings," so that the refreshed booking is stored in the user's bookings list again.

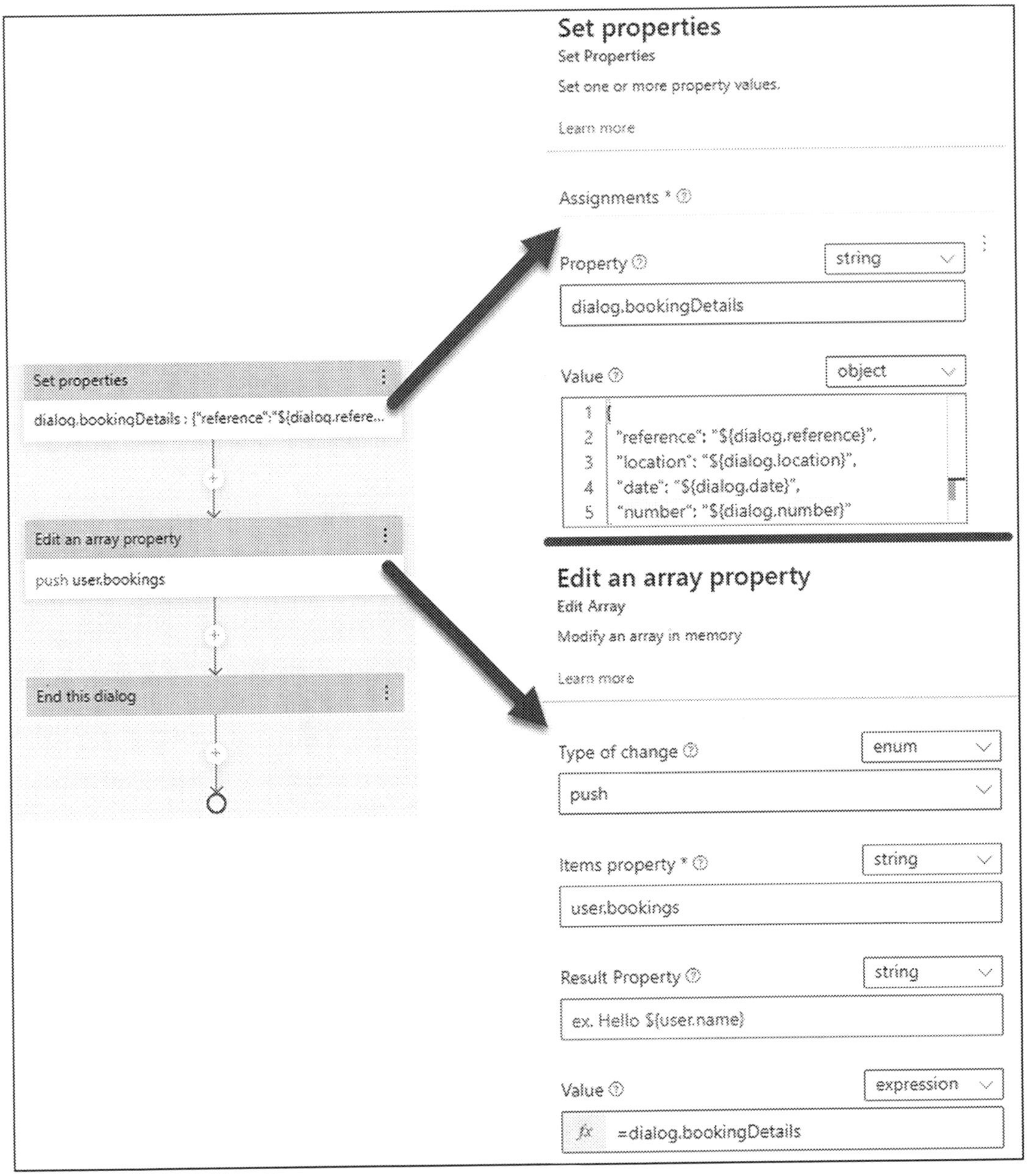

*Figure 5-36.* *ChangeBookingDetails dialog – edit the user's bookings*

Now that the "ChangeBookingDetails" dialog has been finished, the "ManageBooking" dialog will be resumed as outlined in Figure 5-37. The only step left is to ask the user if there is anything else the bot could assist with, to indicate that the bot is also capable of giving further assistance. The user's utterance will again be handled by the top-level LUIS recognizer within the parent dialog in this case.

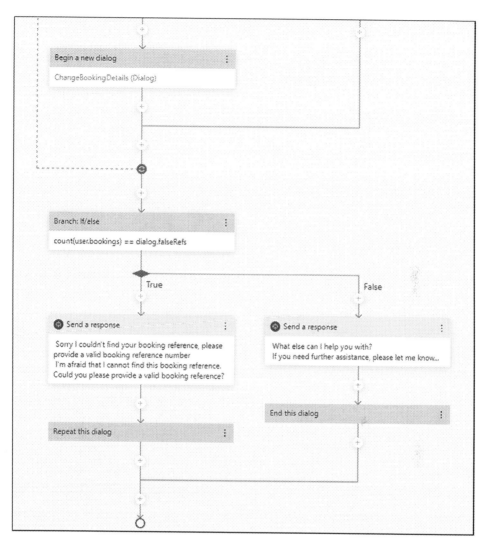

**Figure 5-37.** *ManageBooking dialog – end dialog*

Figure 5-38 summarizes all steps and actions discussed in detail before, as it outlines the overview of the "ManageBooking" dialog basically.

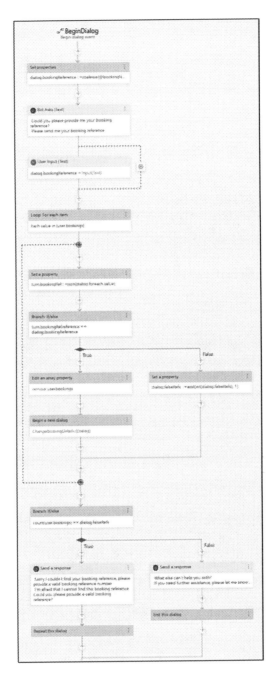

**Figure 5-38.** *ManageBooking dialog – overview*

# Help Dialog

As the users might not always know exactly what they can ask the bot, a "Help" dialog is a good option to describe the bot's capabilities. So, whenever a user types in "Help" or "I need assistance," this dialog will be executed. The "Help" dialog is rather simple, as it consists of two actions as shown in Figure 5-39. The first action is basically used to send an Adaptive Card, with the content to give assistance to a user, to the user. This action makes use of a language generation function "HelpCard." This function basically extracts the "HelpCardJson" string and converts it into a valid JSON payload as well as attaching that payload to an activity in the form of an attachment.

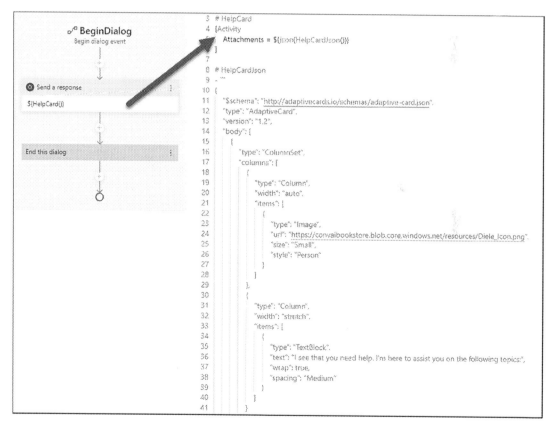

***Figure 5-39.*** *Help dialog*

The outcome of this action is demonstrated in Figure 5-40, which shows that the bot is basically instructing the user how it can help. The Adaptive Card covers all important use cases, which the bot can handle, and which are implemented within the dialogs in Composer. After the card message has been sent out, the dialog will end, and the parent dialog will be resumed. This gives users the option to ask for help within any given turn during a conversation if the flag "Allow Interruptions" is set to true.

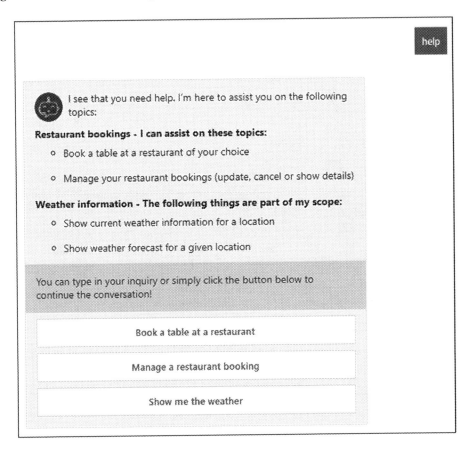

***Figure 5-40.***  *Help dialog – demonstration*

# QnA Dialog

The QnA dialog is used to answer basic or general questions about the booking process for instance. To integrate QnA Maker into a Composer-based chatbot, the easiest way is to add an additional trigger within the root dialog of type "QnA intent recognized," as shown in Figure 5-41. This will make sure that whenever LUIS extracts an utterance and marks it as a QnA Maker–specific question, this dialog will be triggered.

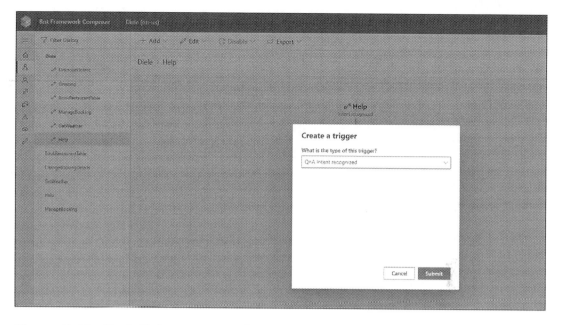

***Figure 5-41.*** *QnA dialog – create trigger*

After adding this new trigger to your bot, the QnA Maker dialog will be automatically added to your bot as shown in Figure 5-42. This prebuilt dialog will basically send the answer which has been received by the QnA Maker endpoint. Optionally, the follow-up prompts will also be displayed as buttons if the QnA Maker answer includes follow-up prompts.

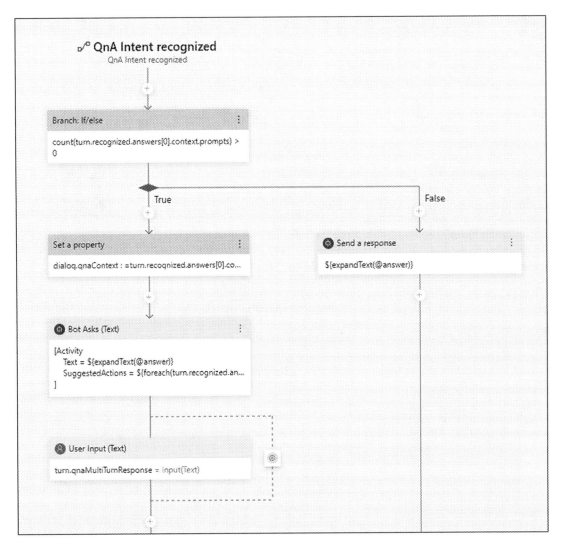

*Figure 5-42.* *QnA dialog – send answers from QnA Maker KB*

One important step to complete, when using QnA Maker in a Composer bot, is to populate the QnA Maker all-up view with questions and answers. Therefore, simply navigate to the QnA Maker all-up view either via the QnA intent trigger as shown in Figure 5-43 or via the QnA button in the navigation menu on the left. Within this view, you can add or alter QnA pairs as you would within the QnA Maker portal, without the need of switching to another portal. When running the bot, which will be described in a later step, Composer creates a new knowledge base within the QnA Maker service you choose automatically and adds the QnA pairs to this new KB as well.

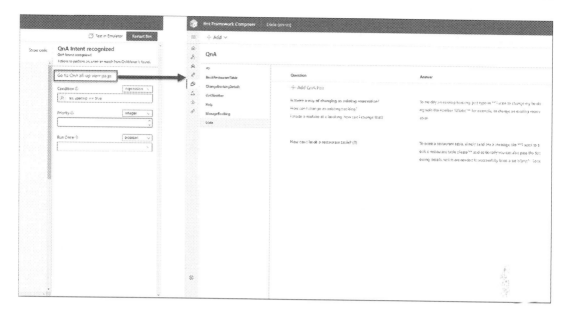

***Figure 5-43.*** *QnA Maker all-up view*

In case you wonder how the complete QnA Maker dialog looks like, Figure 5-44 shows a complete picture of the QnA dialog along with all steps and actions taken when the QnA intent has been recognized by the bot's recognizer.

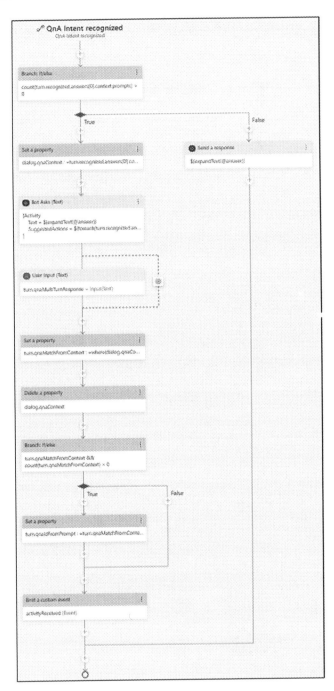

**Figure 5-44.** *QnA dialog – overview*

# Integrate Third-Party APIs into Your Chatbot

A common use case of a bot is not only to handle questions or fulfill specific tasks but also to integrate third-party APIs into the conversation. This is extremely helpful to infuse information from another system directly into the conversation. With this capability, the user does not need to switch context between the conversation and another system. Therefore, the user saves time as the information which would need to be gathered manually in another system can be included in the conversation by connecting the bot with the third-party system. Bot Framework Composer offers a functionality to call an HTTP endpoint at any point within the conversation to either trigger a process in another tool or receive certain information which are then used directly within the conversation with the user. In the sample bot you built so far, a common way which would ease the use of this solution would be to include weather information to let people know the weather forecast before booking a table at a restaurant for instance.

Therefore, the "GetWeather" dialog is used to gather information about the weather using a third-party web API to collect the current weather information as well as the weather forecast of a given city. The very first step is somewhat similar to the other dialogs authored earlier, as you need to collect the information about the city the user wants to get the weather information for. This is done by using the "Set properties" action as displayed in Figure 5-45 which will check if the LUIS recognizer has already extracted the entity "city" from the user's utterance if the "GetWeather" intent has been detected. If the city has not yet been extracted by the recognizer, the bot should ask the user to provide the city name before querying the weather API and store the value in the property "dialog.city."

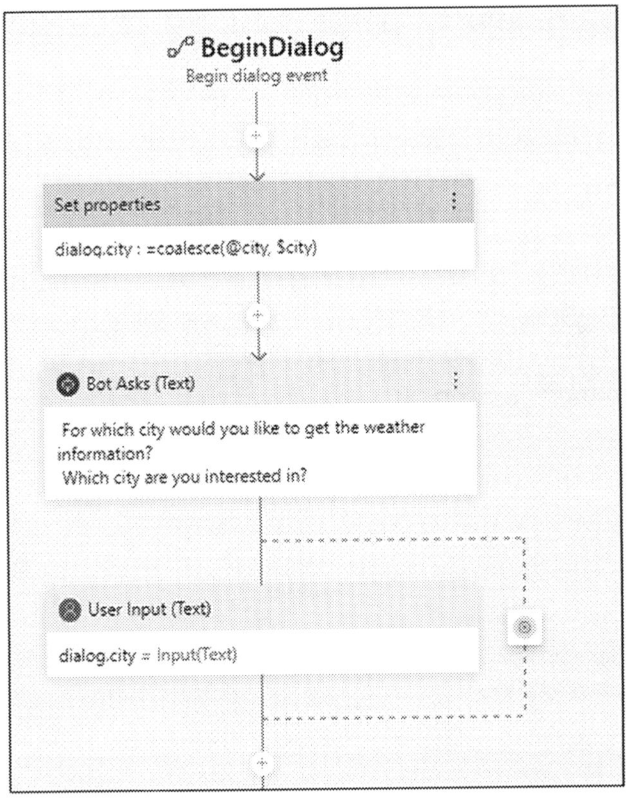

**Figure 5-45.** *GetWeather dialog – set properties and collect information*

After the city name has been set correctly, the next step would be to ask the user if the current weather or the weather forecast should be sent into the chat. This can also be done by the "Bot Asks (Choice)" action to provide two options "Current weather" and "Weather forecast" which the user can choose from as shown in Figure 5-46. The result of that choice prompt will be stored in the "dialog.weatherChoice" property which will be used to determine which API endpoint to use in a later step.

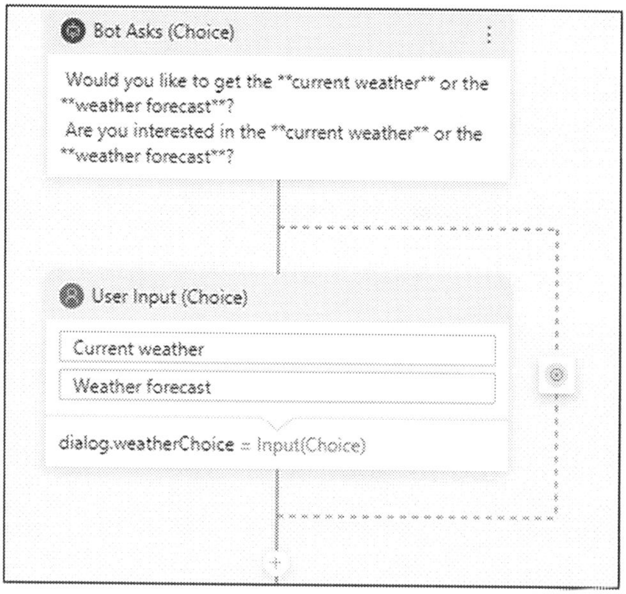

**Figure 5-46.** *GetWeather dialog – choice prompt*

Based on the property "dialog.weatherChoice," the API endpoint for the current weather or the endpoint for the weather forecast will be used in the HTTP request. As demonstrated in Figure 5-47, the "Send an HTTP request" action allows you to choose from the various HTTP methods like *GET*, *POST*, or *DELETE*. Furthermore, the URL can also consist of properties, as seen in the following. This allows you to include properties' values in HTTP requests either in the URL, the headers, or bodies.

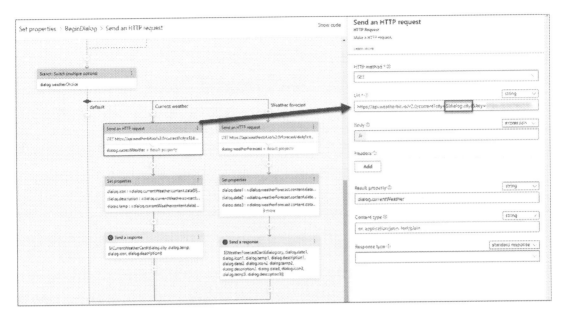

***Figure 5-47.*** *GetWeather dialog – send an HTTP request*

The result of the HTTP request is stored in the "dialog.currentWeather" property. This result property basically holds the complete response information including the HTTP status code, headers, and the content itself. To access the content, simply use "dialog.currentWeather.content" which then holds the content of the API response, so in this case the current weather data. After setting the necessary properties like the icon for displaying the weather, the description, and the temperature, which have been received by the weather API, another LG function "CurrentWeatherCard" will be called in a "Send a response" action, as seen in Figure 5-48.

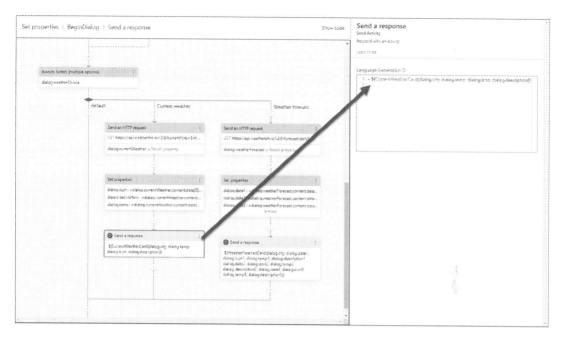

**Figure 5-48.** *GetWeather dialog – send an Adaptive Card response with properties*

This function basically accepts input parameters which are then used to fill in certain values in a predefined Adaptive Card. The following code snippet is basically what you need to use to create this callable function in a language generation file of your bot:

```
# CurrentWeatherCard(city, temp, icon, description)
[Activity
    Attachments = ${json(CurrentWeatherCardJson(city, temp, icon,
    description))}
]
```

As you can see, this function calls another function "CurrentWeatherCardJson" which also accepts the same input parameters as "CurrentWeatherCard" the city, the temperature, the icon, and the description. This function basically contains the JSON payload of the Adaptive Card which should be sent if a user wants to get the information about the current weather. As you can see in the following code snippet, the parameters

are automatically filled in by the language generation engine, if they are following the
${paramaterName} (e.g., *${city}*) notation:

```
# CurrentWeatherCardJson(city, temp, icon, description)
-  ```
{
    "$schema": "http://adaptivecards.io/schemas/adaptive-card.json",
    "type": "AdaptiveCard",
    "version": "1.2",
    "body": [
        {
            "type": "TextBlock",
            "text": "Here is the weather for: ${city}",
            "size": "Large",
            "isSubtle": true
        },
        {
            "type": "ColumnSet",
            "columns": [
                {
                    "type": "Column",
                    "width": "auto",
                    "items": [
                        {
                            "type": "Image",
                            "url": "https://www.weatherbit.io/static/img/
                            icons/${icon}.png",
                            "size": "Small",
                            "altText": "Mostly cloudy weather"
                        }
                    ]
                },
                {
                    "type": "Column",
                    "width": "auto",
                    "items": [
                        {
```

```
                "type": "TextBlock",
                "text": "${temp}",
                "size": "ExtraLarge",
                "spacing": "None"
            }
        ]
    },
...
```

In Figure 5-49, you can see how this dialog would look like in practice. If the user enters a phrases which is identified within the "GetWeather" intent but the LUIS recognizer is not able to extract the city entity from the utterance, then the bot asks back which city the user is interested in. After entering the city name, the user needs to choose between the current weather or the weather forecast. Based on this decision, the bot sends an HTTP request to a third-party API to gather the necessary information and include the details in an Adaptive Card which is then sent back to the user.

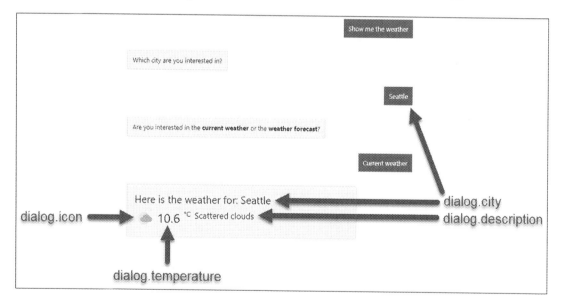

***Figure 5-49.*** *GetWeather dialog – demonstration current weather*

Eventually the dialog could look as the one outlined in Figure 5-50 if the user picks the weather forecast option instead of the current weather information. In this scenario there are multiple instances of the same property type like the icon, temperature, or description as there are multiple days which need to be displayed in the Adaptive Card message. The principle is basically the same as with the current weather approach, although the LG function eventually will look as follows:

```
# WeatherForecastCard(city, date1, icon1, temp1, description1, date2,
icon2, temp2, description2, date3, icon3, temp3, description3)
[Activity
    Attachments = ${json(WeatherForecastCardJson(city, date1, icon1, temp1,
    description1, date2, icon2, temp2, description2, date3, icon3, temp3,
    description3))}
]
```

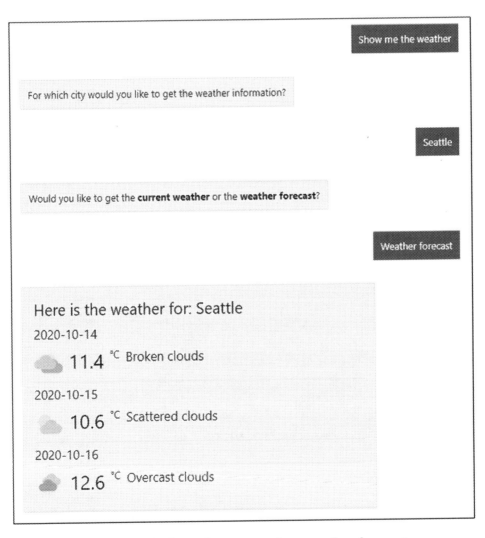

*Figure 5-50.* GetWeather dialog – demonstration weather forecast

An overview of the complete "GetWeather" dialog is outlined in Figure 5-51. In this illustration you see that the API endpoint depends on what the user chooses, either the current weather or the weather information. If the user chooses the weather forecast, then an Adaptive Card with the weather information on the next couple of days will be sent to the user before the dialog ends.

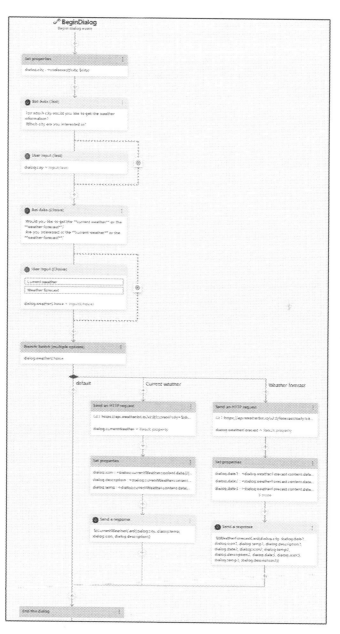

***Figure 5-51.*** *GetWeather dialog – overview*

# Summary

In this chapter, you learned the core concepts about the Microsoft Bot Framework Composer, like how to handle memory management and how to work with language understanding as well as language generation. In addition to that, you built your first bot, which was an echo bot in the first place. This echo bot has then been transformed into a sophisticated chatbot, capable of booking restaurant tables for users within a conversation, managing existing table bookings, or answering questions served from a QnA Maker knowledge base. Moreover, you learned how to integrate a third-party API into the chatbot and therefore enhance your conversation with information gathered from another system to avoid context switches.

The next chapter will be covering the aspect of how to test and debug the chatbot's conversation behavior. You will learn how to use the Microsoft Bot Framework Emulator to test a bot built in Composer as well as how to debug your language understanding application using the emulator.

# CHAPTER 6

# Testing a Chatbot

In Chapter 5, you learned how to build a chatbot using Microsoft Bot Framework Composer. In that chapter you learned how to author the various dialogs and actions to build the conversational experience. The next step after building a chatbot is to test the chatbot. This is the topic of this chapter "Testing a Chatbot" with a strong focus on how to test and debug a chatbot on your machine, without the need to publish the bot beforehand.

During this chapter you will learn how to use the Bot Framework Emulator to test and debug a chatbot's conversation. Furthermore, you will learn how to include tracing events in a Composer dialog to send a trace activity to the Bot Framework Emulator. Additionally, the options to trace the performance of your language understanding service directly within the emulator will be shown.

## Introduction to Bot Framework Emulator

If you are developing chatbots locally, no matter if you are using Composer or the Bot Framework SDK and an IDE to build a bot, you need to test your bot while developing it to see if the conversational experience is as expected. To test the bot's conversation, the Bot Framework Emulator, which is a desktop application, can be used to chat with the bot. In addition, you can also use the emulator to test remote chatbots, which for instance could be hosted in Azure. Besides the capability to chat with the bot, you can also inspect the messages which are exchanged between the user and the bot. The advantage of using the emulator is that you can test and inspect the chatbot locally even if you have not yet published the bot to your Azure subscription.

---

**Note**  To download and install Bot Framework Emulator, visit `http://aka.ms/bfemulator`.

---

© Stephan Bisser 2021
S. Bisser, *Microsoft Conversational AI Platform for Developers*, https://doi.org/10.1007/978-1-4842-6837-7_6

As a bot developer, you should include the usage of the emulator into your development routines. As it is with other GUI-based solutions, the user experience, which is the conversation in a bot's case, is the only thing users will experience. Therefore, the conversational experience should be thoroughly tested before publishing or releasing a chatbot as if the first impression is bad, users will not use your chatbot anymore.

# Testing a Chatbot Using the Emulator

After installing the Bot Framework Emulator, you can connect to any running bot using the bot's messaging endpoint. The messaging endpoint is usually an URL in the following format:

```
http://botEndpoint/api/messages
```

When developing and running a bot locally, this messaging endpoint will most likely be something like this:

```
http://localhost:port/api/messages
```

By default, the port number of a newly created Bot Framework bot is 3979, but this can be changed in the bot's configuration if needed. Therefore, you can run a bot on your local machine either from Visual Studio or via Node.js or other technologies and connect to it within the emulator using the bot URL or messaging endpoint as shown in Figure 6-1. If your bot is already published to Azure or if the application settings of your bot already contain the Microsoft App ID and the Microsoft App password, you need to fill in these values before opening the bot as well. Otherwise the emulator would not be able to connect to your bot successfully, as those parameters are missing.

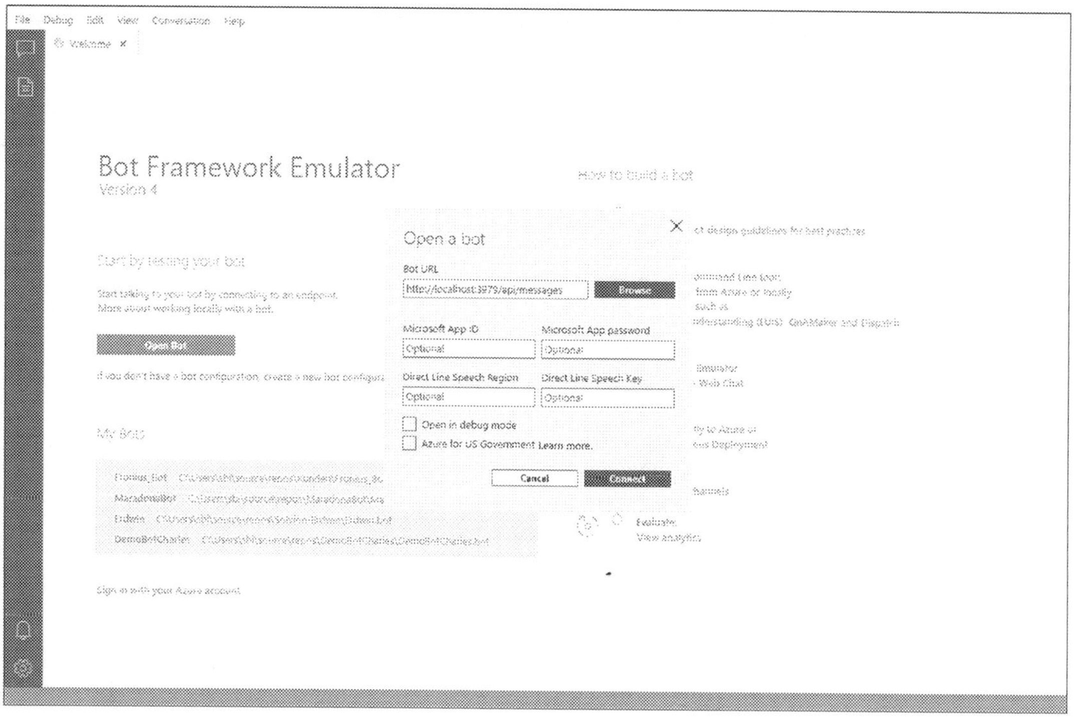

**Figure 6-1.** *Bot Framework Emulator – open a bot*

When building a bot with Composer, you can also start the bot. This process will automatically start the bot on the localhost using .NET core. When starting the bot, the very first time, you need to provide the LUIS authoring key and the QnA Maker subscription key as mentioned in Figure 6-2. The LUIS authoring key is used to publish the LUIS model which has been built using Composer within the dialogs' trigger phrases. The QnA Maker subscription key is used to create and author a QnA Maker knowledge base. The phrases which have been added to the QnA Maker all-up view will be added to the QnA Maker knowledge base. Moreover, if you modify the QnA pairs and restart the bot, Composer will alter the locally made changes in the QnA Maker knowledge base automatically to keep it up-to-date. Furthermore, when starting the bot, Composer will also create a new LUIS application based on the language understanding assets you authored in Composer, like the trigger phrases which contain utterances and entities as well as the intents you defined. Thus, it is not necessary anymore to switch to the LUIS or QnA Maker portal to do some changes within the language understanding model or the knowledge base.

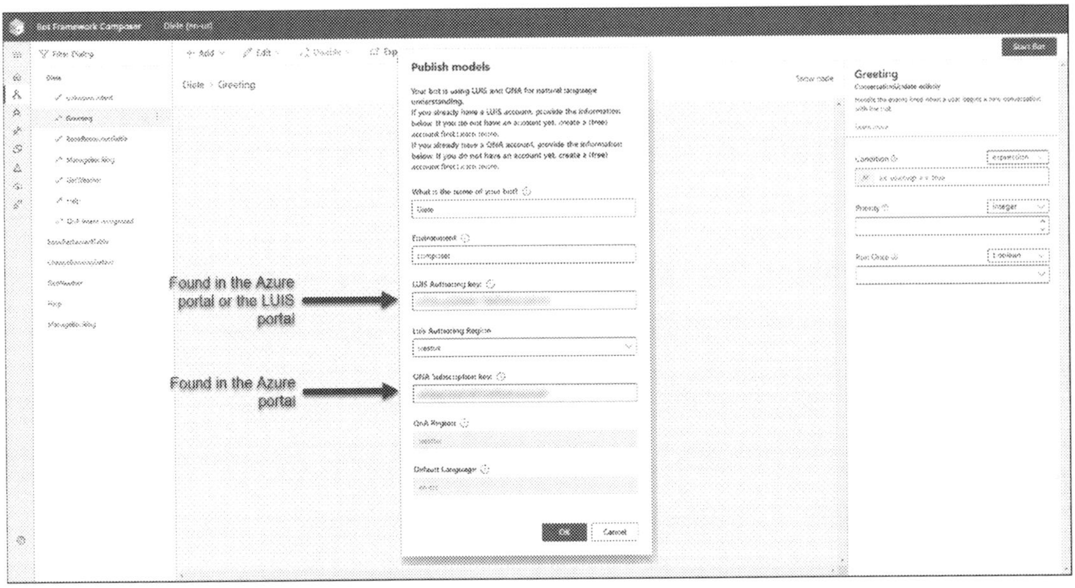

**Figure 6-2.** *Start bot within Composer*

With Bot Framework Composer, this step can be done using Composer. After starting a bot in Composer, a new button will be displayed in the top right corner next to "Restart Bot" which is called "Test in Emulator." This button basically lets you either copy the messaging endpoint URL of the current bot into your clipboard or open the Bot Framework Emulator and connect to the bot directly as shown in Figure 6-3.

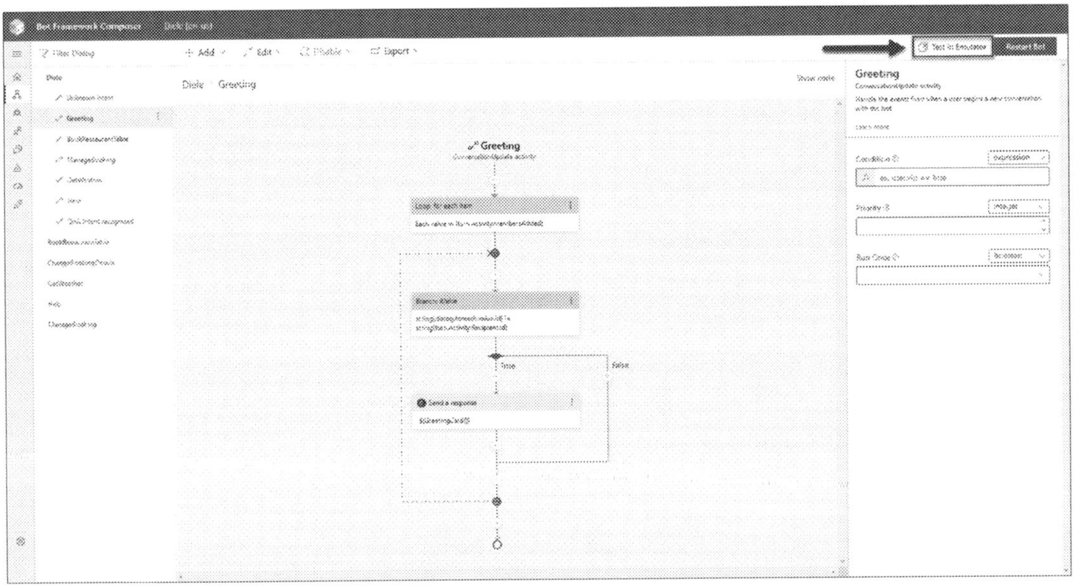

**Figure 6-3.** *Bot Framework Composer – test in emulator*

After the emulator has been opened up and the connection to your chatbot has been established, you can basically have a conversation with your bot similar to users who would use a web chat interface as displayed in Figure 6-4.

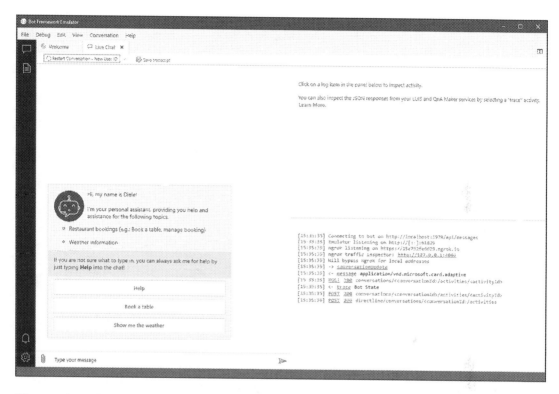

***Figure 6-4.*** *Bot Framework Composer – test in emulator demonstration*

On the left side of the emulator, you basically have a web chat component, which can be used to chat with the bot. The right-hand side, however, offers an inspection section. This inspection panel basically is used to debug and trace the activities which are exchanged between the user and the bot to see what gets sent back and forth. As shown in Figure 6-5, the emulator basically lets you inspect each activity to inspect all the details such as channel data, conversation ID, sender and recipient details, as well as the text and locale of a given message or activity. Therefore, the emulator is basically displaying what the activity consists of behind the scenes.

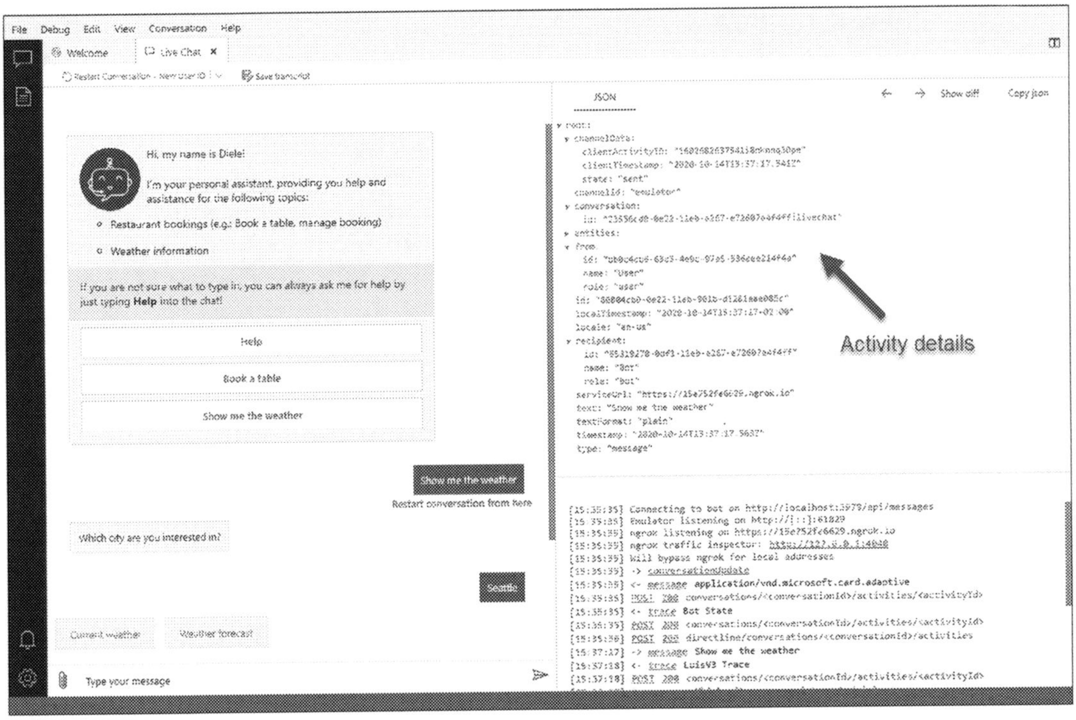

*Figure 6-5.* *Bot Framework Emulator – activity details view*

Composer also offers an action called "Emit a trace event" which can be used to log a new activity into the inspection panel of the emulator. If you for instance want to log the value of a specific property, you can do so by using this action, as shown in Figure 6-6. This is often a helpful situation where you do not want to use the "Send a response" action to log the value of a given property to trace error messages. The users will not notice these kinds of tracing events, so you can also keep them even in production scenarios, as this basically is logging out certain information, which of course should not be sensitive.

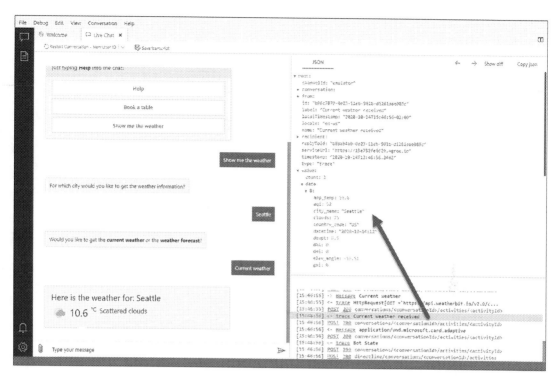

**Figure 6-6.** *Bot Framework Emulator – emit a trace event inspection*

In case of the use case discussed in Chapter 5 when integrating a third-party API and sending an HTTP request, the emulator is also capable of displaying the HTTP request's details, as outlined in Figure 6-7. With this feature you can easily track the HTTP request along with the HTTP method and URL used as well as the HTTP response, including the status code and the response content directly inside the inspection panel of the emulator.

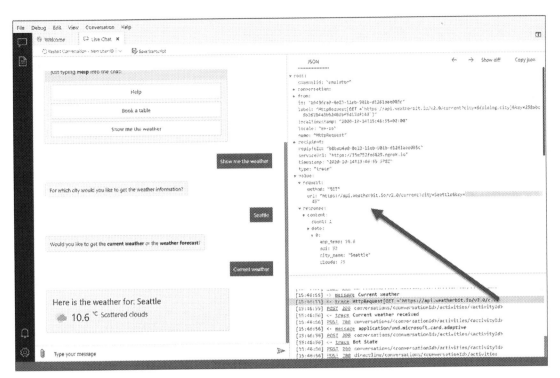

***Figure 6-7.*** *Bot Framework Emulator – inspect HTTP requests*

# Debugging and Tracing Language Understanding

The Bot Framework Emulator is not only intended to debug the conversational behavior of the chatbot. In fact, the emulator can also be used to trace and debug your language understanding model. This can be extremely helpful to test the conversational behavior along with the performance of your language understanding model. This way you can chat with the bot and simultaneously check the detected intents and confidence scores received by your LUIS prediction resource. Moreover, as seen in Figure 6-8, the LUIS trace also includes the extracted entities. This way you can easily track the entities which have been derived from the utterance by the LUIS model. This saves a lot of time when testing and debugging your chatbot, as you can simply test the language understanding model from the emulator. Thus, you do not need to use the LUIS portal to test certain utterances to see what intent will be detected and which entities will be extracted.

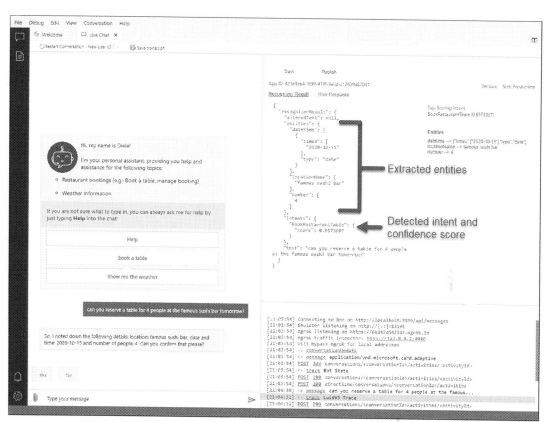

**Figure 6-8.** *Trace LUIS events*

# Debugging and Tracing QnA Maker

The emulator can be used to trace not only a language model but also the QnA Maker behavior. This is helpful in situations where you want to trace the results received by your QnA Maker service, like the confidence scores and most appropriate answers. You can also use the emulator to add alternative phrasing to specific questions or even for training and publishing the QnA Maker knowledge base. Therefore, you do not need to switch context as many common operations can be done using the emulator. To use this QnA Maker trace feature, you unfortunately do some manual preparation beforehand. As this feature is dependent on a .bot file, you need to create and save a new bot configuration file from the emulator on your local machine, as demonstrated in Figure 6-9.

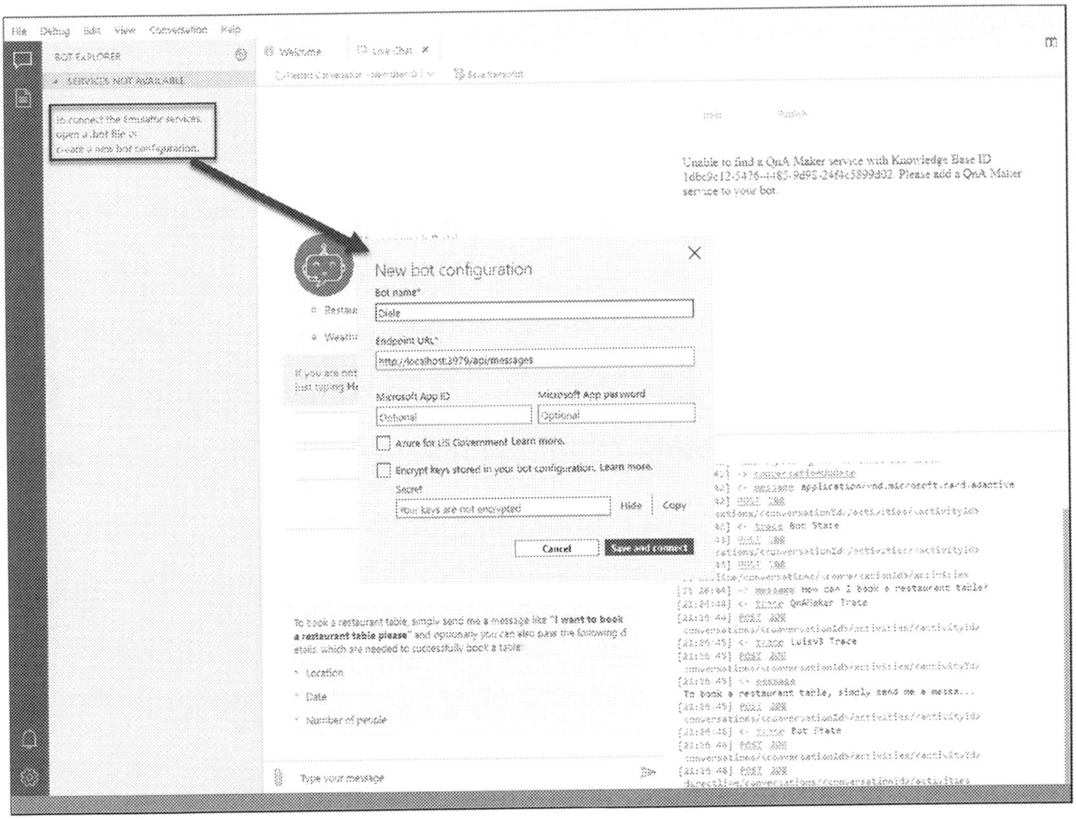

**Figure 6-9.** *QnA Maker trace – create bot configuration file*

After creating the .bot file, the next step before you can use this feature is to sign in with Azure in the emulator. This basically establishes a connection between your Azure account and the emulator, allowing you to connect the emulator to your Azure services directly. Therefore, you need to go to *"File – Sign in with Azure"* (1) and sign in with your personal or work or school account which has permissions on your Azure subscription. After you have signed in successfully, you need to click *"+ – Add QnA Maker"* as shown in Figure 6-10 (2) and then select the corresponding QnA Maker knowledge base (3).

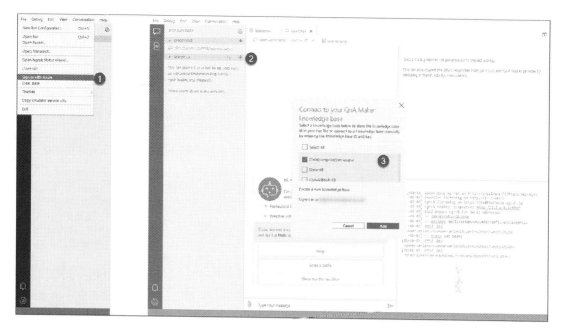

***Figure 6-10.*** *QnA Maker trace – sign in with Azure and connect to QnA Maker service*

After you have added the QnA Maker service to the bot configuration within the emulator, you can go ahead and trace QnA Maker events. To do so simply ask a question which is part of the QnA Maker knowledge base and click the QnA Maker trace event as outlined in Figure 6-11. In there you will see that you receive the option to add alternative phrasing to the given question. Moreover, the confidence scores for the matching question answer pairs are displayed.

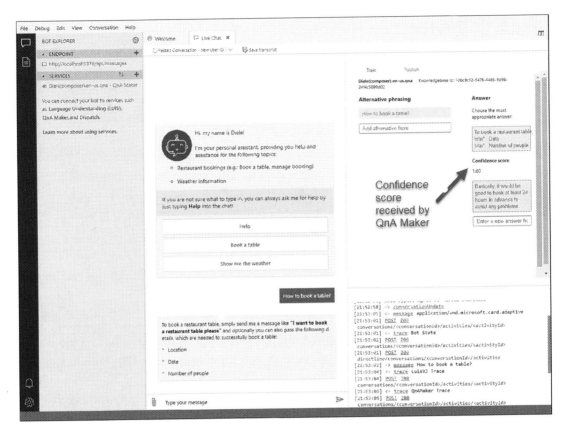

***Figure 6-11.*** *QnA Maker trace – emulator view 01*

Additionally, you will not only see the confidence score of the top ranked question as seen in Figure 6-11. If there are multiple questions which have been matched with the one you entered, you will likely see these as well in the QnA Maker trace, as illustrated in Figure 6-12.

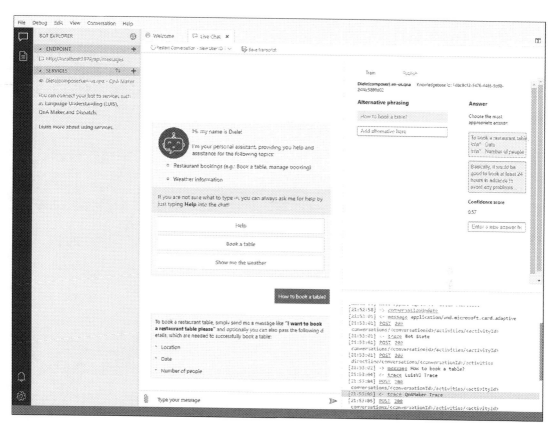

*Figure 6-12.* *QnA Maker trace – emulator view 02*

# Summary

This chapter focused on the process of testing and debugging a chatbot built with Bot Framework Composer. During this chapter you learned about how to use the Bot Framework Emulator to test a chatbot locally, without the need of publishing it to Azure beforehand. Moreover, this chapter covered the process of tracing and debugging LUIS and QnA Maker traces directly within the emulator.

In the next chapter, you will learn about the process of publishing a chatbot built and tested with Composer locally to Azure, which enables you to connect the bot to various channels and release it for your end users afterward.

# CHAPTER 7

# Publishing a Chatbot

Publishing a bot is most likely one of the last steps within your chatbot development life cycle as this operation is done when you built your chatbot and tested it locally until the point you are satisfied with the conversational experience. If this is the case, you can go ahead and publish your bot which enables you to connect it to the channels which are offered by the Azure Bot Service later on, which will be described in Chapter 8. The aim of this chapter, however, is to walk you through the process of publishing a Composer bot to Microsoft Azure.

The publishing process is basically consisting of various steps which need to be executed successfully to have a Composer-based bot up and running in your Azure environment. First of all, you need to install the prerequisites before starting the publishing process:

- Node.js. Use version 12.13.0 or later

- Azure CLI

After installing both prerequisites, you are ready to begin the publishing process of the bot which has been created in Chapter 5.

## Create Azure Resources

Before you can publish the bot's source code to an Azure Web App, you need to provision all necessary resources in Azure like the Azure App Service plan, the Azure Cosmos DB, and the Azure Bot Channels Registration service. To create these Azure resources, you need to open up a terminal instance and navigate to the location where your bot is stored on your local machine. Within this location, there is a folder called "scripts" which you need to navigate to in your terminal window. Within this folder you need to first of all execute the following command to install all dependencies, which are used during the provisioning process later:

```
npm install
```

© Stephan Bisser 2021
S. Bisser, *Microsoft Conversational AI Platform for Developers*, https://doi.org/10.1007/978-1-4842-6837-7_7

After the dependencies have been installed successfully, you can start the provisioning process of your bot. Usually the README file of you bot which is located in the root folder of the bot states that you should run the following command to provision the Azure resources:

```
node provisionComposer.js --subscriptionId=<YOUR AZURE SUBSCRIPTION
ID> --name=<NAME OF YOUR RESOURCE GROUP> --appPassword=<APP PASSWORD>
--environment=<NAME FOR ENVIRONMENT DEFAULT to dev>
```

But the case described throughout this book is slightly different, as we already have provisioned the LUIS service and a QnA Maker service in an Azure resource group already. Therefore, you need to use the following command to provision your Azure resources successfully:

```
node provisionComposer.js --subscriptionId=<YOUR AZURE SUBSCRIPTION ID>
--name=<NAME OF YOUR RESOURCE GROUP> --appPassword=<APP PASSWORD>
--environment=<NAME FOR ENVIRONMENT DEFAULT to dev> --createLuisAuthoring
Resource false --createLuisResource false --createQnAResource false
```

---

**Note**    If you want to provision the Azure services into the already existing Azure resource group, you would need to modify the provisionComposer.js file which is located in the bot's script path to make sure that the script is capable of collecting an existing Azure resource group name as a parameter.

---

After this script has been successfully executed, a new Azure resource group with the name you specified in the preceding command should have been created. This resource group should have the following services included:

- Bot Channels Registration

- Azure Cosmos DB account

- Application Insights

- App Service plan

- App Service

- Storage account

These services are necessary to host your chatbot in Azure. The Bot Channels Registration service is used to connect your bot first of all to the various channels and let the users communicate with the bot via these channels using the Azure Bot Service. The Azure Cosmos DB account is used for the bot's state management. The Application Insights service is used to track the bot's telemetry data as well as the bot's performance, which can be used to generate dashboards in Azure. The App Service plan and the App Service are used to host the bot's runtime. This is basically the web application where the bot's code gets deployed to and which hosts the bot logic as well as the messaging endpoint, which the channels are connected to. The Storage account is used as a transcript store, where all chat transcripts are stored.

The output of the abovementioned script within your terminal window will likely look similar to the one illustrated in Figure 7-1. It is important to copy or store the output from the command somewhere as this is needed in the next step when deploying the bot to Azure.

```
>_ scripts                    ×   +   ∨                                    —   □   ×

> node provisionComposer.js --subscriptionId=f46ecd23-872f-41f2-beb6-bd080777f858 --name=DieleBot --appPassword=cVC
➡!YhzZy7vxAL --environment=dev
Login to Azure:
To sign in, use a web browser to open the page https://microsoft.com/devicelogin and enter the code CM555SRJ9 to au
thenticate.
> Using Tenant ID: 58c5e340-26ff-4bdc-8147-0b98e7074323
> Creating App Registration ...
> Create App Id Success! ID: 12436686-823f-476f-8ea8-ebea1da4b1c3
> Creating resource group ...
> Validating Azure deployment ...
> Deploying Azure services (this could take a while)...
- > Linking Application Insights settings to Bot Service ...
/ > AppInsights AppId: 60dcf4da-298e-47e6-82a6-9126a9fba7a5 ...
> AppInsights InstrumentationKey: dcc647c7-d8a5-466c-bd1a-b4d821d8e609 ...
> AppInsights ApiKey: apv0m0p3zn52242b8lhb8nxcugi44rbs5dfgyh1q ...
\ > Linking Application Insights settings to Bot Service Success!
√ Success!

Your Azure hosting environment has been created! Copy paste the following configuration into a new profile in Compo
ser's Publishing tab.

{
    "accessToken": "
```

*Figure 7-1.* *Provision Azure resources terminal output*

Usually the output of this command may look similar to the following snippet, and
the Azure resource group view should basically look like the one outlined in Figure 7-2.

```
{
    "accessToken": "<SOME VALUE>",
    "name": "<NAME OF YOUR RESOURCE GROUP>",
    "environment": "<ENVIRONMENT>",
    "hostname": "<NAME OF THE HOST>",
    "luisResource": "<NAME OF YOUR LUIS RESOURCE>"
    "settings": {
        "applicationInsights": {
        "InstrumentationKey": "<SOME VALUE>"
        },
        "cosmosDb": {
        "cosmosDBEndpoint": "<SOME VALUE>",
        "authKey": "<SOME VALUE>",
        "databaseId": "botstate-db",
        "collectionId": "botstate-collection",
        "containerId": "botstate-container"
        },
        "blobStorage": {
        "connectionString": "<SOME VALUE>",
        "container": "transcripts"
        },
        "luis": {
        "endpointKey": "<SOME VALUE>",
        "authoringKey": "<SOME VALUE>",
        "region": "westus"
        "endpoint": "https://westus.api.cognitive.microsoft.com/",
        "authoringEndpoint": "https://westus.api.cognitive.microsoft.com/"
        },
        "qna": {
        "endpoint": "<SOME VALUE>",
        "subscriptionKey": "<SOME VALUE>"
        },
```

```
    "MicrosoftAppId": "<SOME VALUE>",
    "MicrosoftAppPassword": "<SOME VALUE>"
  }
}
```

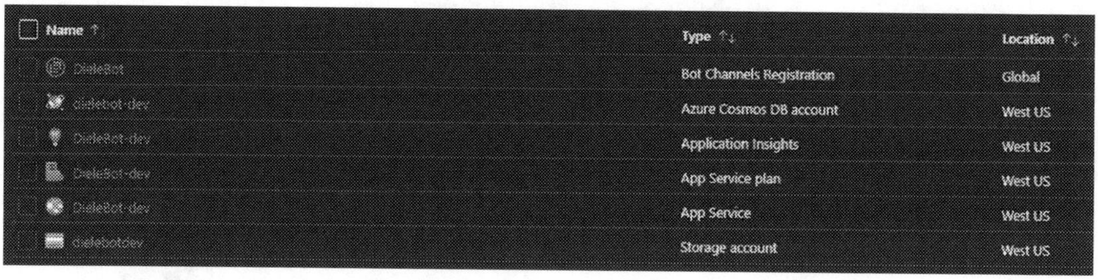

| Name ↑ | Type ↑↓ | Location ↑↓ |
|---|---|---|
| DieleBot | Bot Channels Registration | Global |
| dielebot-dev | Azure Cosmos DB account | West US |
| DieleBot-dev | Application Insights | West US |
| DieleBot-dev | App Service plan | West US |
| DieleBot-dev | App Service | West US |
| dielebotdev | Storage account | West US |

*Figure 7-2.* *Azure resource group after succeeded publishing*

Before using the output mentioned in the preceding text, you need to modify the LUIS and QnA section to include the already existing resources, which will allow the bot to connect to these services after publishing. Therefore, you need to basically change the following keys and properties to the output:

```
{
    ...
    "luisResource": "<NAME OF YOUR LUIS RESOURCE>"
    "settings": {
        ...
        "luis": {
        "endpointKey": "<primary key of prediction resource>",
        "authoringKey": "<primary key of authoring resource>",
        "region": "westus",
        "endpoint": "<endpoint of prediction resource>",
        "authoringEndpoint": "<endpoint of authoring resource>"
        },
        "qna": {
        "endpoint": "<endpoint of QnA Maker KB>",
        "subscriptionKey": "<key of QnA Maker from Azure portal>"
        },
        ...
    }
}
```

The endpointKey as well as the authoringKey and the region can be found in your LUIS app's page under "Manage – Azure Resources." The endpoint and authoringEndpoint can be found in the Azure portal when navigating to the "Keys and Endpoint" page within the LUIS authoring and prediction services. The QnA endpoint and the subscriptionKey can also be gathered from the "Keys and Endpoint" section of your QnA Maker instance in the Azure portal. After altering the output, make sure to copy it as you will need it in the next step for the deployment of your bot.

# Publish Bot to Azure

After the Azure resources have been provisioned successfully, you can go ahead and deploy the bot to Azure. Therefore, you need to switch back to your bot's publish section in Composer as shown in Figure 7-3. This view will most probably not display any information as you have not yet created a publishing profile or published a bot to Azure.

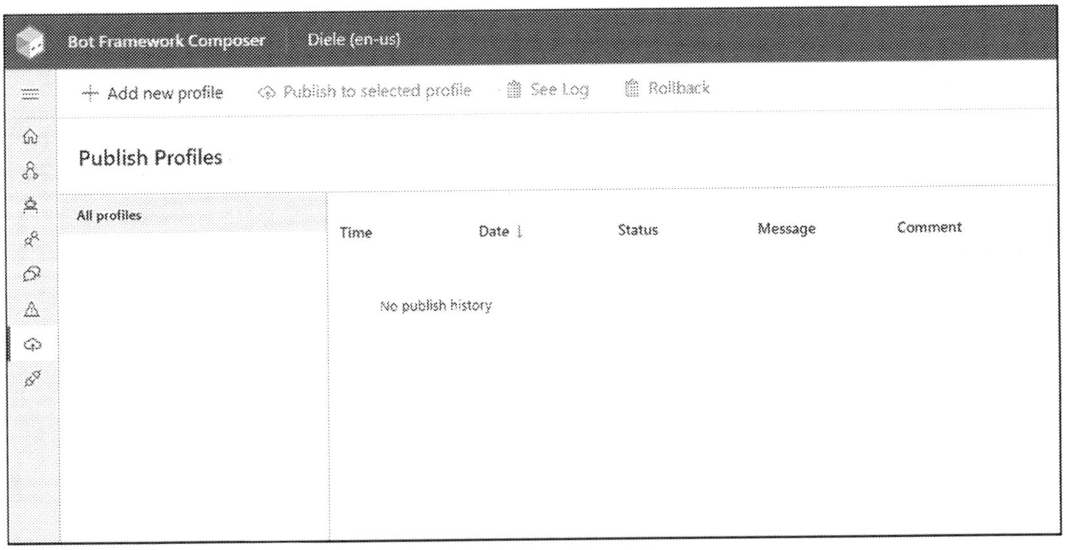

*Figure 7-3. Composer publish view*

Therefore, the first step is to create a new publish profile by clicking the "+ Add new profile" button as shown in Figure 7-4. Then you need to paste in the output from the command within your terminal window into the publish configuration textbox and click save to create the new publish profile accordingly.

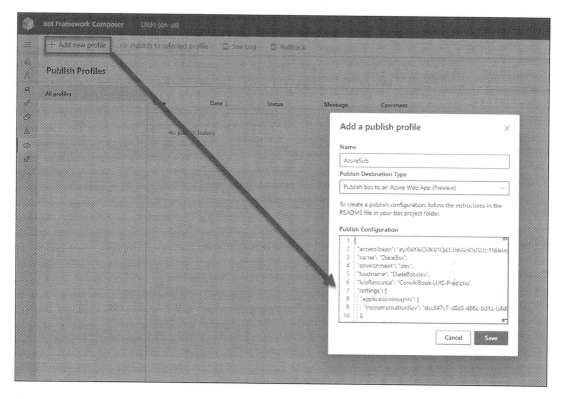

**Figure 7-4.**  *Add new publish profile in Composer*

After the publish profile has been created, you can go ahead and start the publishing process, by selecting "Publish to selected profile" in Composer. Optionally, you can add a comment to this publishing operation, which will be displayed next to the operation later on, as demonstrated in Figure 7-5.

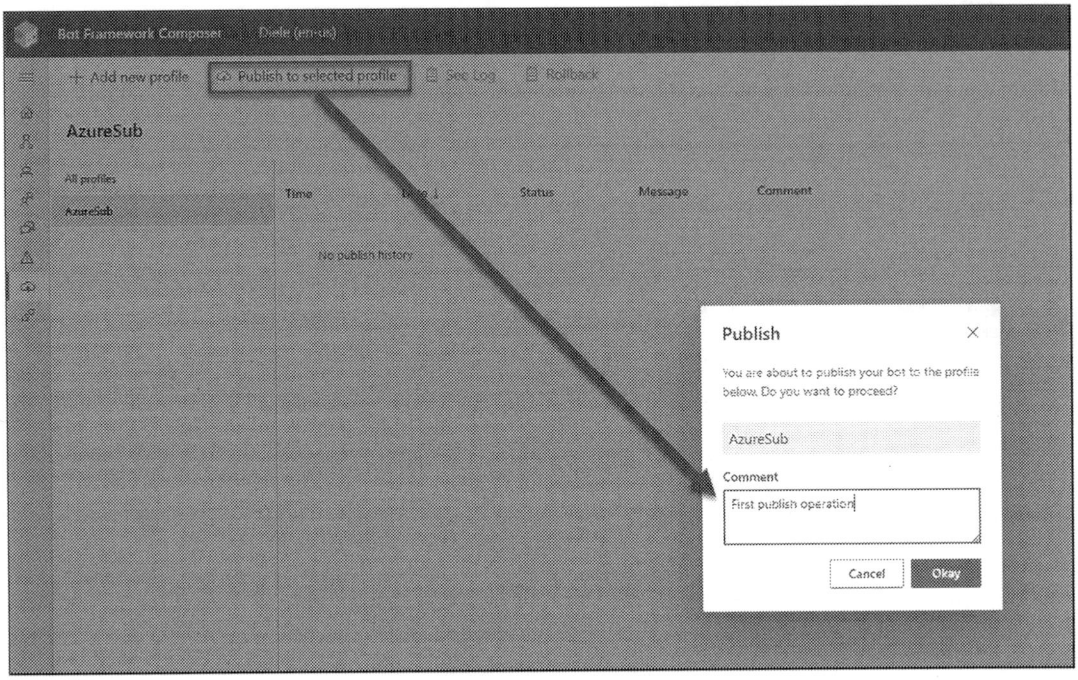

***Figure 7-5.*** *Publish to selected profile*

After the publishing process is finished successfully, you should see message "Success" in your Composer publish view for the newly created publish profile, as shown in Figure 7-6. This basically indicates that Composer was able to successfully publish the bot's code to the Azure Web App resource created before.

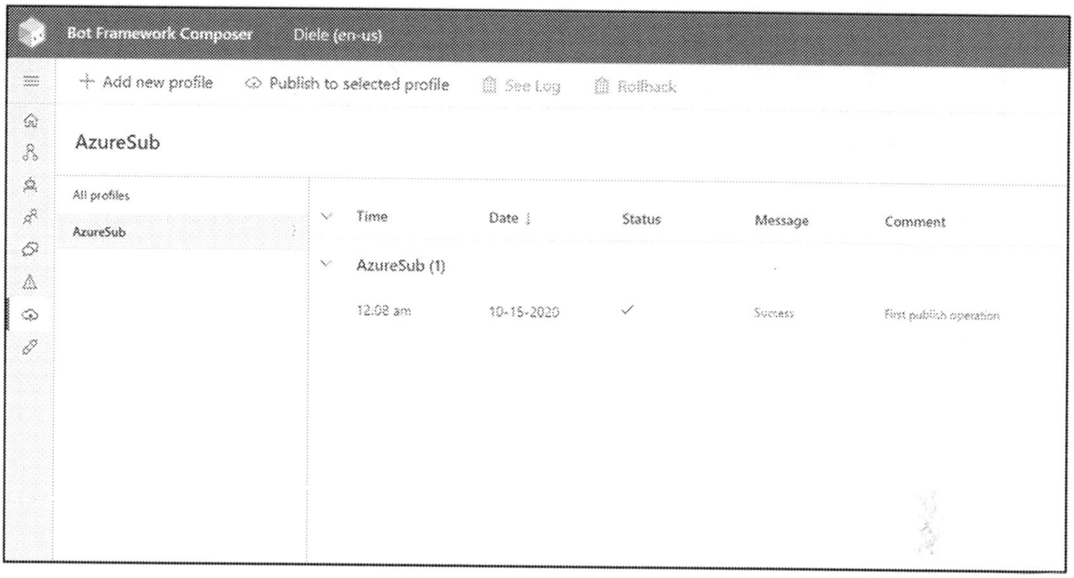

*Figure 7-6.*  *Publish bot success view*

# Test Bot in Azure

Now that the bot has been published to Azure successfully, you can go ahead and test it directly in the Azure portal, similar to the approach used with the emulator. To test a bot in Azure, navigate to the Bot Channels Registration service created in the first part of this chapter. Within that service, navigate to "Test in Web Chat," which will open up your bot in a web chat scenario, as shown in Figure 7-7. If the publish process described earlier has been done successfully, then your bot should greet you the same way, as if you would use the emulator to chat with the bot. Now you can go ahead and test the bot and communicate with it in Azure as well as you would using the emulator.

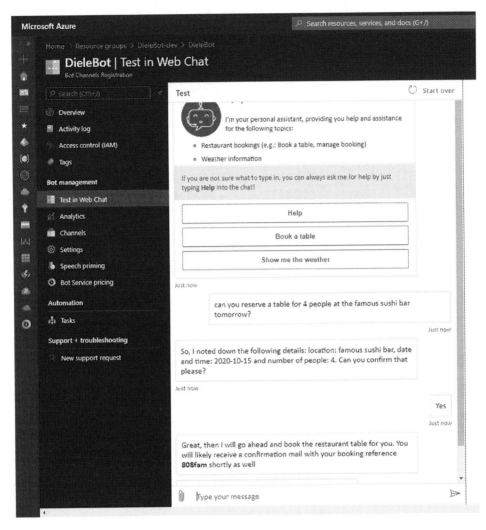

***Figure 7-7.*** *Test bot in Azure after publishing*

## Summary

In this chapter you learned how to publish a chatbot using Bot Framework Composer. In the first part of this chapter, you gained insights in what is required to create the necessary Azure resource needed to successfully publish a bot in Azure. Then, we took look at how to publish a bot using Composer to an Azure subscription as well as test the bot directly from the Azure portal.

The next chapter then deals with the process of connecting the published bot with some of the Azure Bot Service channels like web chat or Microsoft Teams to let end users communicate with the bot.

# CHAPTER 8

# Connecting a Chatbot with Channels

The usually last step in a bot development life cycle is to connect a chatbot to the desired channels. This process basically enables you to let users start communicating with the bot using the channels you enable the bot in. As mentioned in earlier chapters, you should already know which channels you want to target when building your bot as not all channels act the same and support the same features. For instance, Microsoft Teams is capable of handling rich attachments like Adaptive Cards, whereas the Twilio channel is used for text messages. As text messages are not capable of displaying rich attachments, you cannot use Adaptive Cards in this channel. Therefore, you would need to think about the target channels even during the build phase of your development life cycle to avoid major change requests afterward.

The goal of this chapter is describing the supported channels briefly as well as mentioning some differences between some of these channels. Additionally, the process of connecting a bot to the web chat channel and Microsoft Teams is described in detail.

## Azure Bot Service–Supported Channels

As outlined in Chapters 1 and 2 already, the Azure Bot Service is basically the bot hosting platform offered as an Azure service instance. The main feature of this service is to allow bot developers to connect a bot to the supported channels, which are listed here (with subject to change):

- Alexa

- Cortana

- Office 365 email

© Stephan Bisser 2021
S. Bisser, *Microsoft Conversational AI Platform for Developers*, https://doi.org/10.1007/978-1-4842-6837-7_8

- Microsoft Teams

- Skype

- Slack

- Twilio (SMS)

- Facebook Messenger

- Kik Messenger

- GroupMe

- Facebook for Workplace

- LINE

- Telegram

- Web Chat

- Direct Line

- Direct Line Speech

- WeChat

- Webex

The first difference which occurs when comparing the behavior of the various channels is the handling of the welcome message. To get an idea which channels are capable of sending ConversationUpdate activities, please take a look at Table 8-1.

*Table 8-1.* *Welcome message handling by channel (Microsoft, 2020)*

| | Cortana | Direct Line | Direct Line (Web Chat) | E mail | Face book | Group Me | Kik | Teams | Slack | Skype | Telegram | Twilio |
|---|---|---|---|---|---|---|---|---|---|---|---|---|
| **Conversation Update** | 1 | 1 | 1 | 3 | 2 | 1 | 1 | 1 | 1 | 3 | 1 | 3 |
| **Contact Relation Update** | 3 | 3 | 3 | 3 | 3 | 3 | 3 | 3 | 3 | 1 | 3 | 3 |

The keys for the tables mentioned in this chapter are as follows:

- 1 The bot should expect to receive this activity.

- 2 Currently it is undetermined whether the bot can receive this.

- 3 The bot should never expect to receive this activity.

As you can see in Table 8-1, some channels like Teams, Slack, and Telegram support the ConversationUpdate activity which enables you as the bot developer to implement a stable welcome message activity. Bots connected to these channels are capable of sending a greeting message proactively to the user upon starting the conversation. However, other channels like email and Skype are not able to handle these events and can therefore not implement a reliable greeting scenario.

Another example, which shows the differences between channels, is the support of OAuth in a bot connected to the various channels. While some channels like Direct Line or Microsoft Teams support OAuth login, other channels like email do not support this authentication capability. This should also be taken into consideration when building a bot which needs to authenticate users in some sort.

***Table 8-2.*** *OAuth support by channel (Microsoft, 2020)*

| | Cor tana | Direct Line | Direct Line (Web Chat) | E mail | Face book | Group Me | Kik | Teams | Slack | Skype | Tele gram | Twilio |
|---|---|---|---|---|---|---|---|---|---|---|---|---|
| Event. Token Response | 2 | 1 | 1 | 3 | 2 | 2 | 2 | 1 | 2 | 2 | 2 | 2 |

# Connect a Chatbot to Web Chat

As you now learned which channels support which features, it is time to connect the bot you built earlier to the channels offered by the Azure Bot Service. The first channel to connect your bot to is the web chat channel. This channel basically includes the already described web chat control which lets users communicate with the bot within a website or web page. As Figure 8-1 exposes, the web chat channel holds the token used to establish an authenticated connection between the bot exposed in the web chat control and the channel.

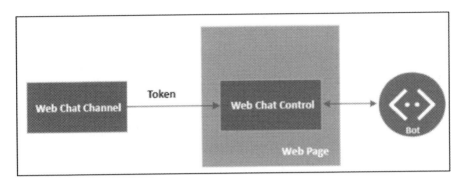

***Figure 8-1.*** *Web chat channel components involved (Microsoft, 2019)*

In general, you have two options how you could embed your bot on a web page using the web chat control:

- Exchange your secret for a token and generate the embed code.

- Embed the web chat control in a website using the secret directly (DO NOT USE IN PRODUCTION).

The first option is the one which provides a by far higher level of security, as your secret is not exposed on your web page as you change the secret for a token and then generate the web chat control embed code automatically. The following HTML code snippet basically outlines how to establish a secured connection without exposing the secret as it is retrieved by an API automatically and is therefore not part of the code:

```
<!DOCTYPE html>
<html>
  <head>
    <script src="https://cdn.botframework.com/botframework-webchat/latest/
    webchat.js"></script>
  </head>
  <body>
    <div id="webchat" role="main"></div>
    <script>
     const styleSet = window.WebChat.createStyleSet({
         bubbleBackground: 'rgba(0, 0, 255, .1)',
         bubbleFromUserBackground: 'rgba(0, 255, 0, .1)',
         botAvatarImage: '<your bot avatar URL>',
         botAvatarInitials: 'BF',
```

```
      userAvatarImage: '<your user avatar URL>',
      userAvatarInitials: 'WC',
      rootHeight: '100%',
      rootWidth: '30%'
    });
    // Calling the API to retreive the token
    const res = await fetch('https:<YOUR_TOKEN_SERVER>', {
        method: 'POST' });
    const { token } = await res.json();
    window.WebChat.renderWebChat(
      {
        directLine: window.WebChat.createDirectLine({ token }),
        userID: 'WebChat_UserId',
        locale: 'en-US',
        username: 'Web Chat User',
        locale: 'en-US',
        styleSet
      },
      document.getElementById('webchat')
    );
  </script>
  </body>
</html>
```

The web chat channel is enabled by default when creating a new bot in Azure. Therefore, you can use the abovementioned HTML snippet to embed your bot in a web page. The Direct Line secret can be gathered from the Channels section of your bot in the Azure portal as Figure 8-2 shows.

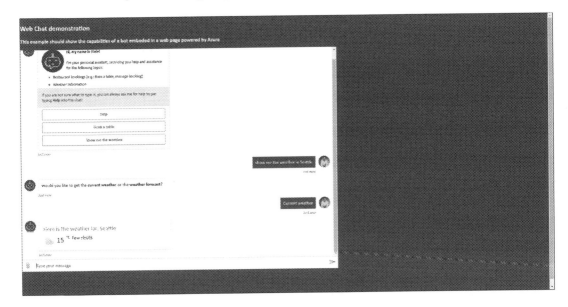

**Figure 8-2.** *Web chat channel configuration*

After embedding the HTML snippet in a web page, the chat window could look as the one shown in Figure 8-3. But of course, as this is just HTML and CSS, you can modify the look and feel to fit your needs or your corporate branding guidelines. You can also create a pop-up chat window on one side of the page which allows users to open the chat window when they are seeking help for instance.

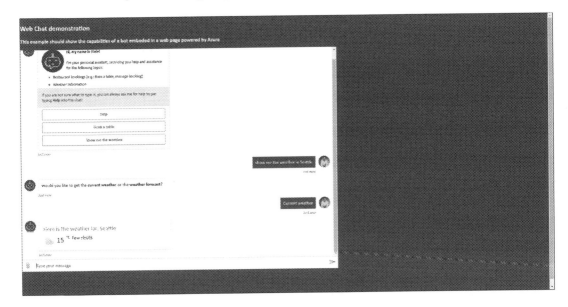

**Figure 8-3.** *Web chat implementation example*

**Note**    If you need more details on how to customize the web chat control, please conduct `https://docs.microsoft.com/en-us/azure/bot-service/bot-builder-webchat-customization`.

# Connect a Chatbot to Microsoft Teams

Another popular channel to connect a bot to is Microsoft Teams, as it is tightly integrated into the modern workplace within Microsoft 365. To connect a bot to Teams, you would first of all need to enable the Teams channel from the Azure portal, by clicking the Microsoft Teams icon as shown in Figure 8-4.

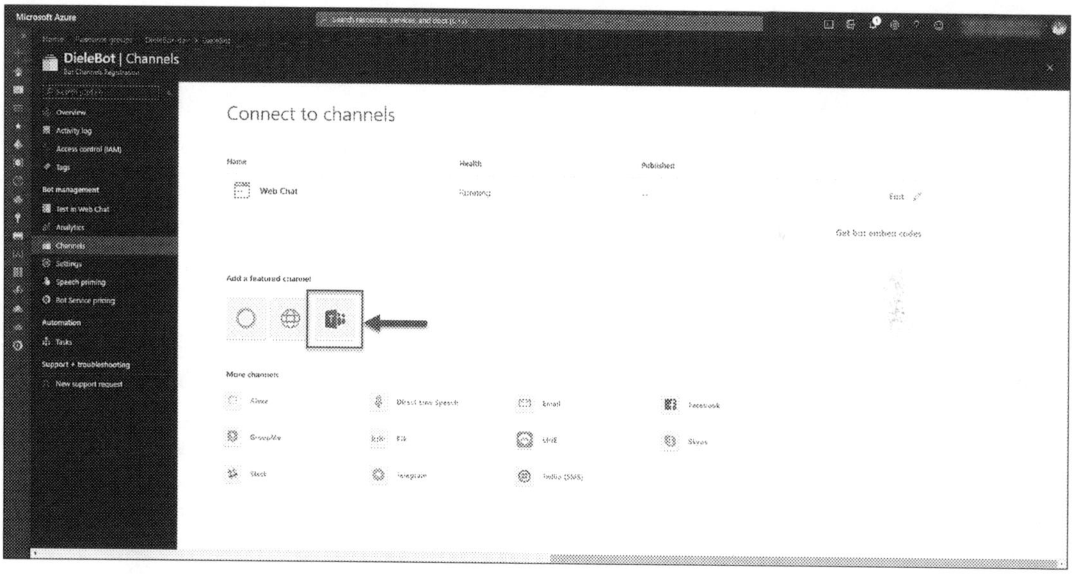

***Figure 8-4.*** *Enable Teams channel 01*

On the next page, you simply need to save the configuration and accept the terms of conditions as outlined in Figure 8-5. This step basically activates the Teams channel for your bot and provides the possibility to chat with the bot via the Microsoft Teams channel.

**Figure 8-5.** *Enable Teams channel 02*

If you want to test if the bot is responding in the Microsoft Teams channel, simply click the Microsoft Teams icon from the list of channels as demonstrated in Figure 8-6. This will open the Microsoft Teams web or desktop client and establish a connection to the bot in the client automatically for you. If everything has been set up correctly, you should see the welcome card a few seconds later sent to you by the bot.

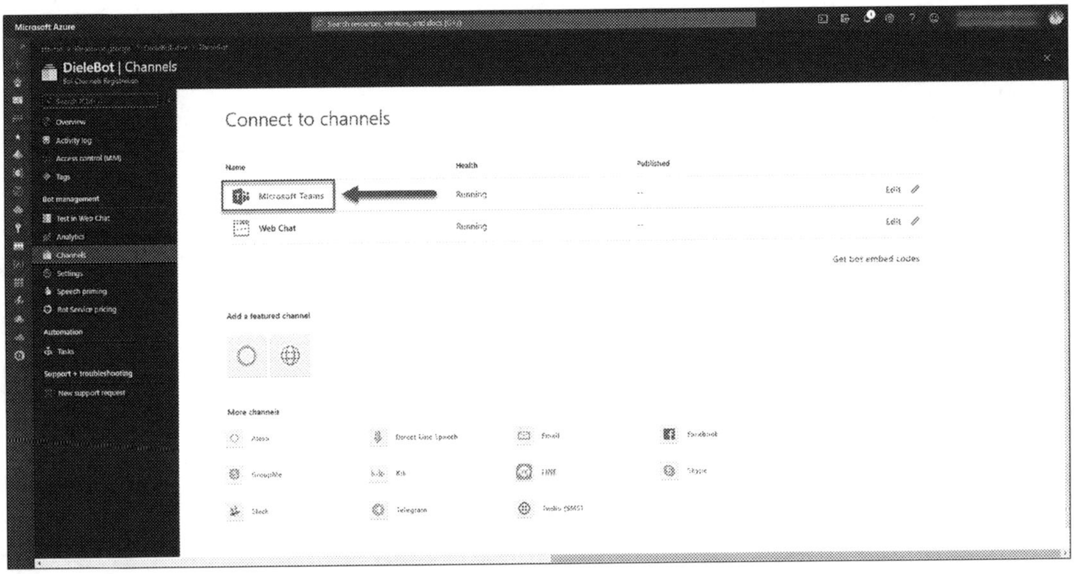

**Figure 8-6.** *Test bot in Teams*

But in production scenarios, you would not likely give access to all people who should be allowed to communicate with the bot via Teams to the Azure portal page or send out the direct URL to the bot in Teams. Therefore, you should create a new Microsoft Teams application using the Teams App Studio, consisting of the bot. To create such a Teams app, open the Teams App Studio and create a new app. The first thing to do is to fill out the basic information about the app such as the name and the developer and optionally include app icons, as shown in Figure 8-7.

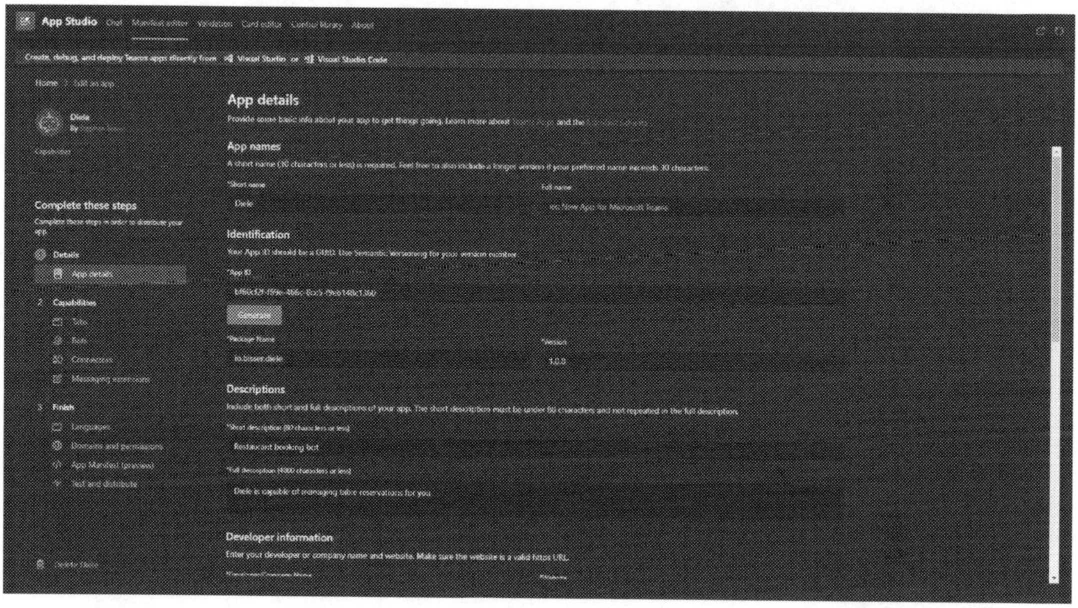

***Figure 8-7.*** *Create Teams app for the bot – app details*

**Note**   Your user needs to be able to access the Teams App Studio, which may be blocked by the administrators of your organization.

After providing the necessary app details, you can switch to the "Bots" section of the app to set up your bot accordingly as outlined in Figure 8-8. This step basically connects your app to the bot hosted in Azure to allow users to communicate with the bot via this app.

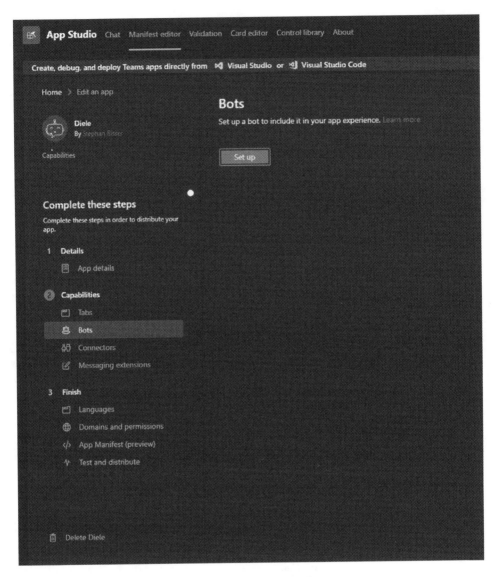

**Figure 8-8.** *Create Teams app for the bot – set up bot 01*

To set up the bot in the Teams app, you need to get your bot's Microsoft App ID, which is called "Bot ID" in the Teams App Studio. This App ID can be found on the settings page of your bot in the Azure portal as exposed in Figure 8-9.

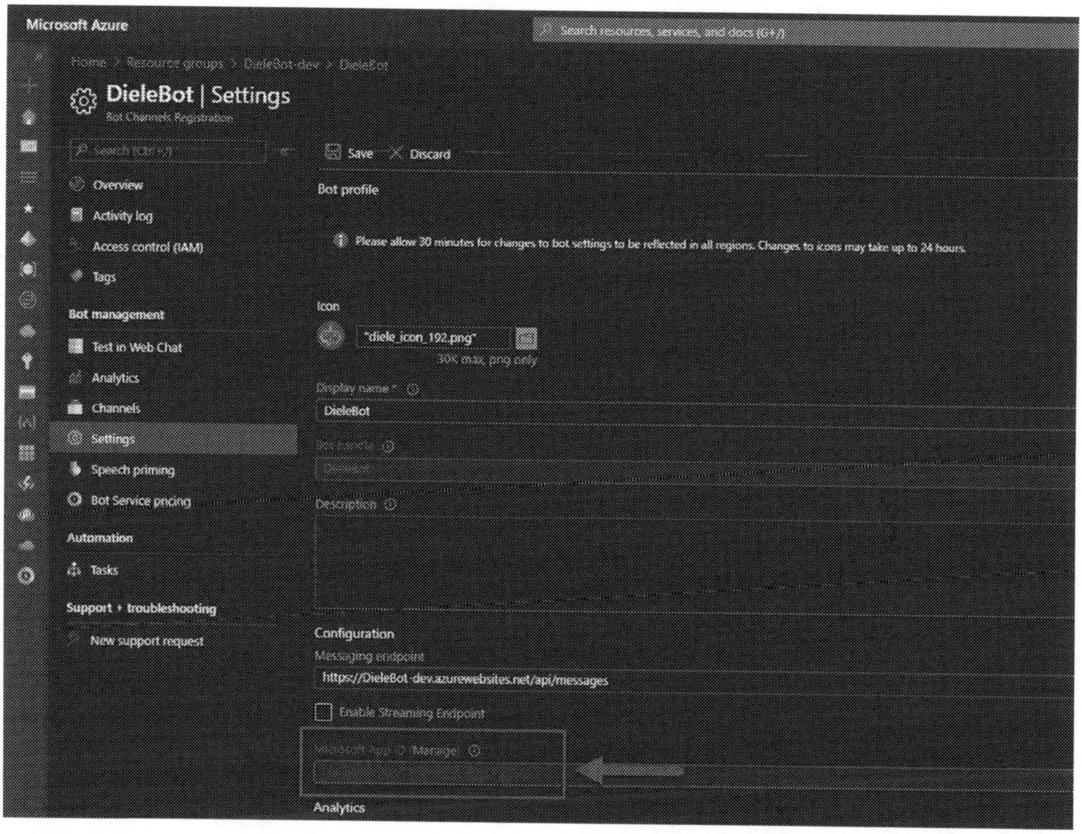

*Figure 8-9.* *Create Teams app for the bot – Microsoft App ID*

To set up a bot within a Teams app, you need to provide the "Bot ID," which is the "Microsoft App ID." Moreover you need to select the bot's scope which can be "Personal," "Team," and "Group" to choose in which scenarios users should be allowed to communicate with the bot. Additionally you can enable the support to upload and download files within the conversation or restrict the bot to be a notification-only bot as outlined in Figure 8-10.

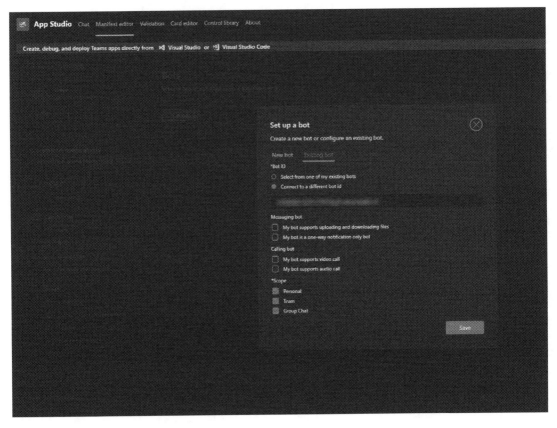

***Figure 8-10.*** *Create Teams app for the bot – set up bot 02*

After the bot has been set up correctly within the Teams app, the next step is to install the bot for you to see if the bot is able to communicate with a user via Teams. Therefore, you need to go to the "Test and distribute" section and then choose "Add for me" to install the bot in your Teams client as an app, as illustrated in Figure 8-11.

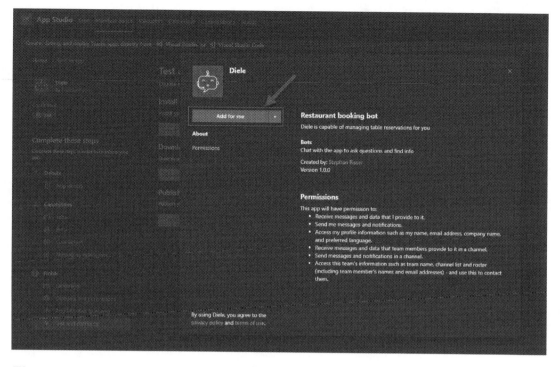

***Figure 8-11.*** *Create Teams app for the bot – install bot in Teams*

After installing the bot, you will see that the bot's icon is displayed in the left navigation bar of your Microsoft Teams client. Furthermore, a few seconds after installing the bot, you should receive the welcome message by the bot which can be seen in Figure 8-12, the same way as you would when using the web chat control or the Bot Framework Emulator to establish a communication with the bot.

***Figure 8-12.*** *Test bot in Teams*

This Microsoft Teams app can now be uploaded to the public Microsoft Teams store to allow everyone to use your bot. If you only want to let people within your organization use your bot in Microsoft Teams, you can upload the Teams app to your organization's app store, so that only members of your organization can install and use your bot.

# Summary

In Chapter 8, we covered the Azure Bot Service–supported channels. In particular, we compared certain features across the channels to see which channels support which key features. Moreover, you learned how to connect a bot to the web chat channel and integrate the web chat control securely into a web page without potential security problems. Additionally, we demonstrated how to customize the web chat control and apply styling to include a corporate design even within the conversation experience. The last section of this chapter walked you through the process of enabling the Microsoft Teams channel for your bot as well as creating and installing a Microsoft Teams app containing your bot to distribute across your organization.

# Index

## A

Activity event triggers, 157

Artificial intelligence, *see*
    Conversational AI

Azure Bot Service (ABS), 11

## B

BookRestaurantTable dialog
    ask a question action, 209
    confirmation option, 211–212
    end dialog action, 213
    overview, 214
    prompt feature, 210
    set properties, 207–215

Bot framework composer
    advantages/disadvantages, 178
    differences, 178
    features/capabilities, 177
    language generation
        (LG), 184–186
    language understanding
        (LU), 182–184
    memory scopes
        accessibility/durability, 179
        explicit type *vs.* expression, 181
        expressions, 180
        properties, 180–182
    options, 177
    overview, 178

Bot framework service
    activities, 26
    adaptive cards, 59–62
    blob storage, 32
    command-line interface, 58
    composer
        adaptive dialogs, 64
        advantages, 63
        integrate language generation, 65
        language understanding, 64
        overview, 62–63
    conversation, 26
    conversation state, 33
    Cosmos DB storage, 32
    dialogs/prompt types, 34–37
    echo bot project, 39–46
    emulator, 53–54
    handlers, 28–30
    hosting platform, 51–52
    HTTP POST, 27
    incoming activity, 27
    JavaScript, 30, 46–50
    key concepts, 25
    memory storage, 32
    middleware processing concept, 37–39
    multiturn conversations, 31
    private conversation state, 33
    proactive messaging, 27
    processing, 27–28
    skills reusable components, 50, 51

© Stephan Bisser 2021
S. Bisser, *Microsoft Conversational AI Platform for Developers*, https://doi.org/10.1007/978-1-4842-6837-7

Printed in the United States
By Bookmasters